Quarter Chassis Technology

By Steve Smith

Photos by Steve Smith and Peter Twill

Cover by Lorianne Twill

ISBN 0-936834-54-4

© Copyright 2004 by Steven R. Smith. No part of this publication may be reproduced, stored in a retrieval system or transmitted, in any form or by any means, electronic, mechanical, photocopying, recording or otherwise, without the prior written permission of the Publisher.

Published By

STEVE SMITH AUTOSPORTS® PUBLICATIONS

P.O. Box 11631 / Santa Ana, CA 92711 / (714) 639-7681
www.SteveSmithAutosports.com

Printed and bound in the United States of America

Table of Contents

Introduction 4
   Thank You 4
   Disclaimer Notice 4

1. Race Car Handling Basics 5
   Handling 5
   Total Traction Capacity 5
   Understeer and Oversteer 6
   Lateral Weight Transfer 7
   Lateral Weight Transfer Effect on Traction 8
   Rate of Weight Transfer During Cornering 9
   Proportioning Weight Transfer 10
   Roll Couple Distribution 10
   Lateral Acceleration 11
   Center of Gravity 12
   Roll Centers 12
   The Roll Moment 13

2. Suspension Systems 15
   Front Suspension 15
   Caster 15
   Camber 17
   Kingpin Inclination 18
   Scrub Radius 19
   Ackerman Steering 20
   Anti-Roll Bar 22
   Rear Suspension 23
   Four Link Suspension 24
   Unlocked/Locked Left Rear 25
   Left Rear Ratchet Hub 26
   Left Rear Coil-Over Location 26
   Roll Centers 26
   Front Roll Center 28
   Rear Roll Center 29
   Spring Rates 29
   Coil-Overs 30
   Sprung Vs. Unsprung Weight 30
   Weight Adjustment 31

3. Shock Absorbers 33
   The Purpose of Shocks 33
   Double Tube, Low Gas Pressure Shocks 33
   Monotube High Gas Pressure Shocks 34
   Shock Absorbers and Handling 35
   How Shocks Influence Handling 36
   Shock Damping Stiffness 37
   Damping Stiffness Codes 38
   Split Valving Shocks 38
   How Shocks Influence Handling 39
   Chassis Tuning With Shocks 40
   Loose At Corner Entry 40
   Tight At Corner Entry 40
   Loose At Mid Corner 40
   Tight At Mid Corner 41
   Loose At Corner Exit 41
   Tight At Corner Exit 41
   Other Shock Tuning Tips 42
   Dirt Track Conditions and Shock Requirements .42
   Heavy/Tacky Track 42
   Dry Slick Track 42

4. Tires & Wheels 43
   Tire Dimensional Sizing 43
   Tire Compounds 43
   Racing Tire Durometer Hardness Chart 44
   Tire Heat Cycles 45
   Tire Softeners & Chemical Treatment 46
   Track Tire Rules 46
   Tires For Qualifying 46
   Tire Inflation Pressure 47
   Dirt Track Air Pressure 48
   Factors Affecting Tire Pressure 49
   Using Nitrogen 49
   Checking Tire Pressure 50
   Checking Tire Temperatures 50
   Tire Stagger 51
   Growing A Tire 51
   Reading Tire Surfaces 51
   Racing Tire Break-In Procedure 53
   Tire Use And Aging 54
   Tire Mounting 54
   Tire Selection For Dirt Tracks 55
   Selecting The Right Dirt Tire Compound 56
   Wheels 56
   Wheel Backspacing 57
   The Effects of Wheel Width 57
   One-Piece Vs. Two-Piece Wheels ...58
   Wheel Inspection 58

5. Chassis Setup & Alignment 59
   The Baseline Chassis Setup 59
   Square The Rear Axle 59
   Adjusting The Rear Axle To Make It Square 60
   Leveling The Birdcages 60
   Squaring The Front Axle 62
   Setting the Panhard Bars 62
   Setting Caster 63
   Setting Camber 64
   Checking Toe-Out/Toe-In 65
   Get the Car Race-Ready 66
   Adding Weight for A Class 67
   Chain Tension 68

Setting Ride Heights .69
Ride Height Setting Shortcut 70
Scaling the Chassis 71
Scaling the Car After A Race 76
Front Axle Lead 76
Rear Axle Lag 77
Choosing Spring Rates 77

6. Track Tuning & Adjustment 79
Chassis Setup Philosophy 79
Watch the Driving Line 79
Chassis Sorting at the Track 80
What Adjustments Are Available and What Do They Do? 81
Stagger 81
Wheel Spacing or Tracking 83
Tire Pressure 85
Adjusting Tire Pressures 86
Rear Panhard Bar 86
Front Axle Lead 87
Chassis Tilt 88
Rake 88
Cross Weight 89
Shock Absorber Valving 89
Spring Rates 89
Tire Compounds 90
Toe Out 90

Chassis Tuning With Weight Jacking 91
The Unresponsive Chassis 91
Handling Problems & Troubleshooting 92
Adjusting For A Push 92
Car Is Loose At Corner Exit 92
Bicycling 93
Tuning For More Grip 93
Tuning For Less Grip 93
Chassis Tuning With Tire Temperatures 94
Chassis Adjustment Quick Reference 96
Adjusting the Chassis For A Dirt Track 96
Slick Dirt Track 96
Sticky Dirt Track 96
Reading A Dirt Track .96
Chassis Adjustment Quick Reference 97
Dry Slick Track Adjustments 98
Sticky Track Adjustments 98
Chassis Changes Required For Dirt Tracks 99
Spring Rates 99
Tracking 99
Tilt 100
Tire Pressure on Dirt 100
Chassis Ride Height 100

Chassis Setup Sheet .101

Suppliers Directory 102

Thanks

No book of this size and scope could ever be written without the help and cooperation of many people. A very special thanks goes to all of the people whose ideas and information helped contribute to this book. Thank you to John Nervo of NC Chassis Co., and Ron Stanley of Robbie Stanley Racing. And, a very special thank you goes to Bill Carlson, a very knowledgeable and successful handler, for sharing his knowledge and experience.

Most importantly, a very special thank you goes to my very understanding and dedicated family who have to make special allowances for me while I am researching and writing a book. Thanks for your support!

Steve Smith

Introduction

Many people entering quarter midget racing are new to the sport of auto racing. They find the world of chassis tuning and adjustment very perplexing.

This book was written to help the competitors in this sport to better understand the intricacies of quarter midget handling basics, terminology, set up procedures, and preparation.

Our goal in writing this book is to make life easier for the racer...to help him understand his race car so that racing can be an enjoyable, instead of frustrating, experience. Hopefully this book will fulfill that goal for you.

Have fun, be safe, and happy racing!

Steve Smith

Disclaimer Notice

Every attempt has been made to present the information contained in this book in a true, accurate and complete form. The information was prepared with the best information that could be obtained. However, auto racing is a dangerous undertaking and no responsibility can be taken by any persons associated with this book, the authors, the publisher, the parent corporation, or any person or persons associated therewith, for injury sustained as a result of or in spite of following the suggestions or procedures offered herein. All recommendations are made without any guarantee on the part of the author or the publisher, and any information utilized by the reader is done so strictly at the reader's own risk. Because the use of information contained in this book is beyond the control of the author or publisher, liability for use is expressly disclaimed.

We recognize that some words, model names and designations mentioned herein are the property of the trademark holder. We use them for identification purposes. This is not an official publication.

Chapter 1

Race Car Handling Basics

Handling

Everything that happens to a vehicle in motion (at least while it is on the ground and upright) happens at the tire contact patch. Those four little patches of rubber that the car sets upon are ultimately what we must deal with. The whole idea is to make these contact patches generate as much force as possible. The entire scope of handling comes down to this simple concept.

The following information is intended to increase your knowledge of basic race car dynamics – the forces that affect the performance of a race car, and ultimately, affect force generation at the contact patches.

Everything that happens to a vehicle in motion happens at the tire contact patch. The concept of maximizing handling is to make these contact patches generate as much force as possible.

Total Traction Capacity

There are a number of factors which determine the traction capacity of a vehicle. The construction characteristics and rubber compound of the tires are the most obvious. Tire contact patch area is another. Suspension geometry will determine how well the contact patches stay in contact with the track. Gross vehicle weight is a prime factor (less weight being better). The height of the Center of Gravity above the ground and the way the vehicle transfers weight, both laterally and longitudinally, affect the tire vertical load. Lateral means side-to-side, and longitudinal means front-to-rear or rear-to-front.

Most of these factors are difficult to change, as they are determined in the design stage of the car. A brief summary of these design characteristics include:

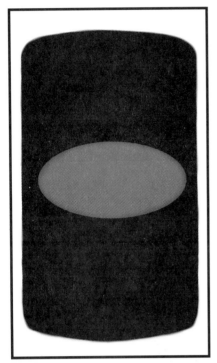

A race car handles best when the entire width of the tire contact patch is equally loaded. Loading to one side or just the center takes away from the total traction capability of the tire.

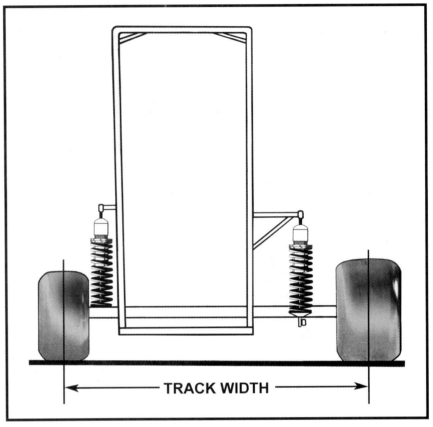

Track width is the distance from the center of the left tire to the center of the right tire on the same end of the car.

1. Gross vehicle weight. Most classes have a minimum weight rule. Any given car should be at or below the minimum weight, or at least be equal to the lightest cars running in that class. Cars below minimum weight can add ballast to improve the weight distribution as well as bring the car up to minimum weight requirement.

2. Center Of Gravity location. Center of gravity is a factor in determining lateral weight transfer during cornering and longitudinal weight transfer during acceleration.

3. Track width. Track width is the distance from the center of the left tire to the center of the right tire on the same end of the car. This is usually controlled by rules, but the maximum is best as this is also a factor in lateral weight transfer. On dirt tracks, depending on the track condition, track width may be narrowed to influence or improve side bite (tire grip that limits outward sliding of the tire and race car).

Understeer and Oversteer

Understeer and oversteer are the two basic reactions of a vehicle when steering forces direct a car off its straight heading. Understeer is a resistance to any steering force requiring that more steering wheel angle be added to stay on the radius of the turn. In racing language, understeer is pushing or plowing of the front end of the vehicle.

Oversteer is just the opposite of understeer. It means that the rear end of the vehicle is loose, requiring less and less steering wheel angle to keep the vehicle on the radius of the turn.

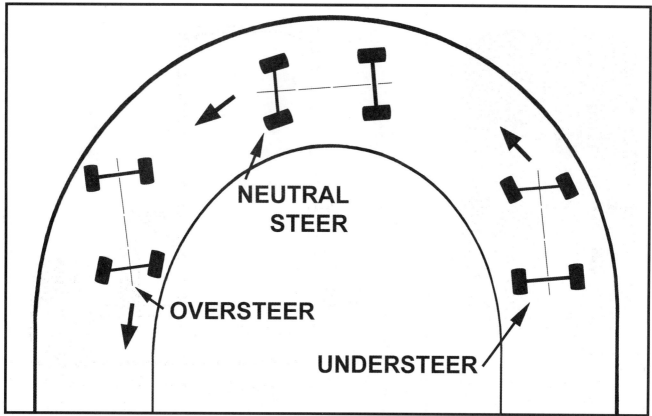

Neutral steer is the path the car takes when it stays on the proper driving line without sliding the front or the rear tires. Oversteer (also called "loose") and understeer (also called "pushing") both cause tire scrub, which costs substantial lap time because the driver has to lift off the throttle to regain traction.

Understeer and oversteer can be "transient" or momentary conditions, and both conditions may be experienced while traversing a single turn. A car which has a basic oversteering handling characteristic may still understeer while entering a turn because of some transient handling characteristic.

There are a number of factors which have an influence on understeer and oversteer. Two of the most important are weight distribution of the vehicle and roll couple distribution (which is explained later in this chapter). Having a basic design influence on the two above factors are front and rear roll center locations and center of gravity height.

These basic influences all come into play when the vehicle experiences lateral acceleration and weight transfer during cornering.

Lateral Weight Transfer

Any time a race car drives through a corner, the load on the left side tires is reduced and the load on the right side tires in increased by a like amount. This is called "weight transfer". There are three major factors which affect weight transfer: the car's weight, center of gravity height and its track width. The formula for finding the amount of weight transfer is:

$$\text{Transfer} = \frac{W \times CGH \times G}{TW}$$

In other words, the amount of weight transfer in pounds equals the car weight (expressed in pounds) times the center of gravity height (expressed in inches) times the

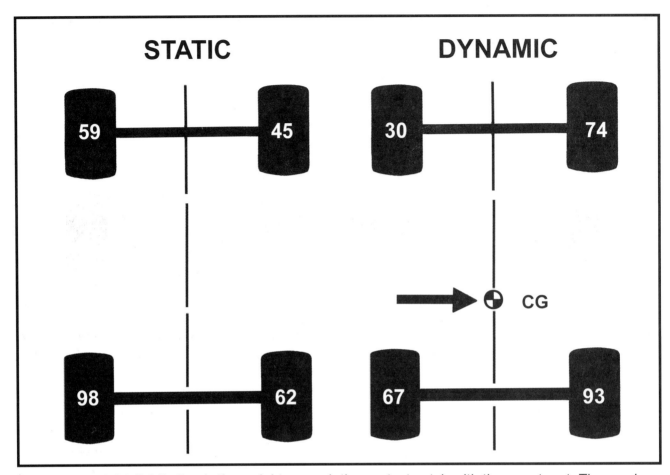

The static weight distribution is the weight on each tire contact patch with the car at rest. The numbers shown above are actual numbers of a 260-pound class (with driver). During cornering, which is the dynamic state, weight transfers from the inside to the outside tires. Centrifugal force acts against the car's center of gravity (CG), causing the chassis to roll about the roll axis.

lateral force coefficient (expressed in g's) all divided by the track width of the car (expressed in inches).

A vehicle with no suspension will transfer the same amount of weight laterally as a car with suspension if these three factors remain the same.

Lateral Weight Transfer Effect on Traction

When weight is transferred during cornering, the vertical load on each tire is changed, and this affects the traction of each tire. Tire traction is increased as vertical loading on it is increased. But this relationship is not linear. As loading is placed on a tire, the traction increases at a slower rate than the rate of weight increase. This same principle applies in reverse as loading is taken off a tire. Traction is decreased as loading on a tire is decreased, but the loss of traction occurs at a faster rate than the weight decrease. These facts are true for all tires, but it will vary for individual tires due to tire design and inflation pressure.

Traction is best when a pair of tires on an axle is equally loaded. Weight transfer during cornering decreases this equality of loading, and thus traction ability, because the outside tires gain more traction while the inside tires are losing traction at a faster rate than the outside tires are gain-

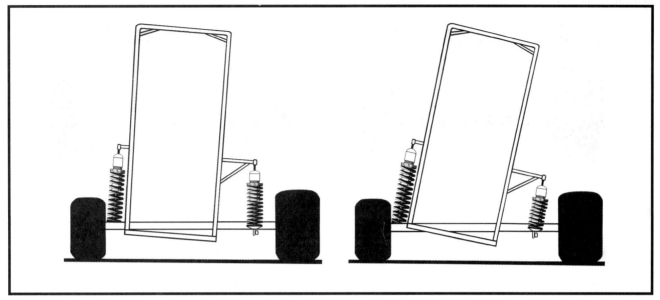

If a car uses stiffer rate springs (left), less body roll will occur. The same amount of weight will transfer from inside to outside, but less chassis roll will occur because the stiffer outside springs will have more resistance. If a car uses softer rate springs (right), more body roll will occur. This is because the weight being transferred will compress the softer springs more.

ing traction.

To better equalize tire loading during cornering, start with more static (at rest) weight loading on the left side tires. The total amount of weight transfer will remain the same during cornering, but increasing the left side static weight means more weight will remain on the left side tires during cornering. This improves tire traction and cornering speed. The goal for a quarter midget chassis setup on a paved track application is 57 to 60 percent left side weight.

Rate of Weight Transfer During Cornering

Springs and shock absorbers do not control how much weight is transferred during cornering. They merely influence what happens to the chassis during the transfer process.

Weight transfer is determined by lateral acceleration, the center of gravity height, and the track width of the vehicle. No matter what the spring rates are on the car, or how stiff or soft the shocks are, the car will transfer the same amount of weight with all the factors in the weight transfer formula remaining the same. If you change the car's track width, or lateral G force or CGH, then the amount of weight transferred will change.

Spring rates affect two things during the weight transfer process: 1) how much chassis roll the car has, and 2) how the transferred weight is proportioned between the right front and right rear wheels.

If a car uses stiffer rate springs, less body roll will occur. The same amount of weight will transfer from inside to outside, but less chassis roll will occur because the stiffer outside springs will have more resistance. If a car uses softer rate springs, more body roll will occur. This is because the weight being transferred will compress the softer springs more.

Shock absorbers, as well, do not influence the amount of weight being transferred. A shock absorber is a valved hydraulic device that resists motion. By resisting suspen-

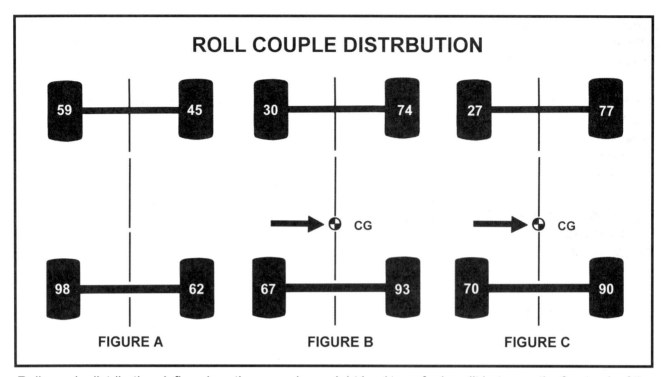

Roll couple distribution defines how the cornering weight load transfer is split between the front pair of tires and the rear pair of tires. **Figure A** *shows the car with the static (at rest) corner weight distribution.* **Figure B** *shows dynamic cornering with 49% front roll couple, which means 49% of the overturning weight transfer occurs at the front wheels, and 51% of the load transfer takes place at the rear axle.* **Figure C** *shows dynamic cornering with a 52% front roll couple. Notice that more weight is transferred at the front, and less at the rear. The front weight transfer increase resulted from an increase in front spring rates. This makes more of the lateral weight transfer occur at the front.*

sion movement, a shock can influence the rate at which weight transfer and chassis roll occur. The rate is how fast the movement occurs. A stiffly valved shock will offer more resistance, and thus slow down the rate at which weight transfer occurs. A lightly valved shock will offer less resistance to movement, so it allows weight to transfer at a faster rate.

Proportioning Weight Transfer

Proportioning the weight transfer between the front tires and rear tires is what race car handling adjustment is all about. More (or excessive) weight transfer at the front tires will cause a car to understeer (also called pushing or tight). More (or excessive) weight transfer at the rear tires will cause a car to oversteer or be loose.

The goal, using chassis tuning tools such as spring rates or roll center heights or corner weights or wheel tracking, is to balance the front and rear weight transfer to neutralize handling and prevent pushing or loose conditions. The goal is to equalize tire grip on the right front and right rear tires during all phases of cornering.

The formal name for proportioning the front-to-rear weight transfer during cornering is called roll couple distribution.

Roll Couple Distribution

Roll couple is the force, due to cornering, acting on the sprung weight of the vehicle, rolling it about the roll axis. It is the action that transfers weight from the inside wheels to the outside wheels.

The front-to-rear handling characteristics of a race car can be tailored or adjusted by adjusting the front-to-rear roll stiffness proportioning. This is called roll couple distribution. It is how the weight transfer is distributed between the front suspension and rear suspension when chassis roll occurs. The end of the car with the highest roll stiffness will receive the greatest amount of weight transfer during body roll.

The resistance created by the front and rear suspension systems can be adjusted by spring rates to change the weight transfer distribution between the front and rear of the car. For example, increasing a right front spring rate increases the front roll stiffness of the car, and thus increases the front roll couple distribution. A stiffer right front spring makes more of the cornering weight transfer be handled at the front end of the car.

Roll couple is computed by first determining the wheel rate of each of the car's four springs. These rates are all added together, giving sum number one. Then add the wheel rates of the front springs together to get sum number two. Divide sum number two (the total front rate) by sum number one (the total vehicle rate) to obtain the front roll couple percentage. The rear roll couple percentage is determined by subtracting the front roll couple percentage from 100 percent.

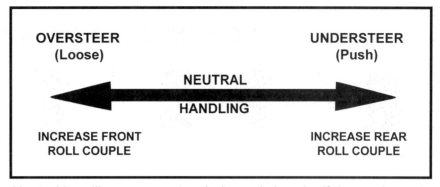

Neutral handling represents a balanced chassis. If the car is pushing or tight (understeer), increase the rear roll couple to balance it. If the car is loose (oversteer), increase the front roll couple to balance it.

The higher the front roll couple percentage, the stiffer the front roll stiffness is, which means more of a tendency toward understeer. Conversely, the less front roll couple, the more tendency toward oversteer. The more rear weight percentage a particular car has, the stiffer the front roll couple must be in order to balance the tire slip angles on the car.

Lateral Acceleration

Lateral acceleration is the sideways force which a car generates in a turn. It is a measurement of the maximum amount of cornering force a car will generate. This lateral acceleration is closely related to centrifugal force, which is a force that makes a rotating body move outward from the center of rotation. The difference is that lateral acceleration is expressed in terms of "G's" of force whereas centrifugal force is measurable in actual pounds of force. Lateral acceleration is dependent on two things: velocity (speed) and the radius of a turn. The formula to find lateral acceleration is:

$$LA = V^2 / 32R$$

Or, in other words, lateral acceleration is the velocity squared (expressed in feet per second) divided by the radius of the turn times 32. Centrifugal force is the lateral acceleration multiplied by the total weight of the car.

Once the lateral acceleration coefficient for a particular car is known, the actual pounds of weight a car is transferring from the inside to the outside wheels during maximum cornering can be calculated. This is explained in the Lateral Weight Transfer section in this chapter.

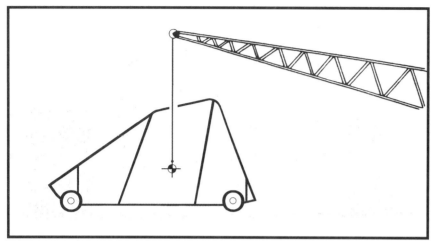

If you could position a crane so it could hook the actual center of gravity of the car, it would suspend the car perfectly balanced. You would be able to rotate it 360 degrees around the hook and it would remain perfectly balanced.

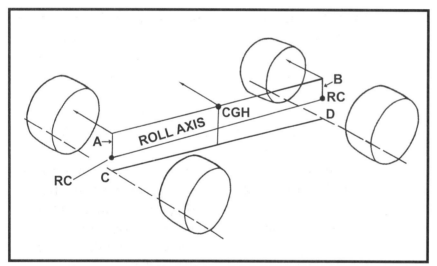

The relationship between the roll centers, roll axis and CGH. "A" is the front roll center moment arm. "B" is the rear roll center moment arm. Line "CD" is the ground plane. As the CGH rolls about the roll axis, weight is transferred from the inside to the outside.

Center of Gravity

The center of gravity of a car is an imaginary point which is the absolute center of all weight in the car — vertical, front to rear, and left to right.

If you could position a crane so that it could hook the actual center of gravity, it would pull the car up off the ground perfectly balanced. You would be able to rotate the vehicle 360 degrees around the hook, and the car would remain perfectly balanced. This would demonstrate that the CG is the actual physical center point of weight mass balance in the car.

The center of gravity height (CGH) is the balance point in the chassis which evenly splits the upper and lower weight masses of the car. The CGH is important because it is that point through which the centrifugal force acts during cornering. The higher the CGH is above the roll axis, the more weight that will transfer from inside to outside during cornering.

There are also two other CG's — the fore/aft center of gravity, which is the balance point halfway between the front and rear weight masses, and the left/right CG, which is the center of the left and right weight masses.

Roll Centers

Roll centers (front and rear) are two of the most important design parameters in the race car. Together with the center of gravity height, these parameters influence everything the car does, the way the car behaves, and the effects of suspension changes you make to the chassis. Too often, a racer just accepts whatever front and rear roll centers occur in the car through the placement of components, without realizing the effect of the roll centers on the total handling characteristics during cornering.

The effect of centrifugal force during cornering acts on the total car mass through its center of gravity. This force is experienced as "lean" or "sway," but most commonly known in vehicle dynamics as "roll." The effect of this centrifugal force on the center of gravity is totally dependent on speed (velocity). The higher the center of gravity, the lower the speed at which a car will reach its maximum roll.

When the body rolls, just like anything in a circular motion, it must have a central point about which it rotates, or rolls. In a car this is known as the roll center.

In any car there are two roll centers — front and rear — and they are determined totally independent of each other. But when a car is cornering, the front and rear suspensions are connected through the car's chassis and they both experience roll at the same time. So an imaginary line called a roll axis connects the front and rear roll centers.

Roll centers are determined completely by suspension system geometry. Quarter midget race cars use a beam or straight axle suspension system both in the front and rear suspension system. This axle is attached laterally to the chassis with a Panhard bar. When suspension lateral movement is controlled with a Panhard bar, the roll center is located at the center line of the Panhard bar, both laterally and vertically.

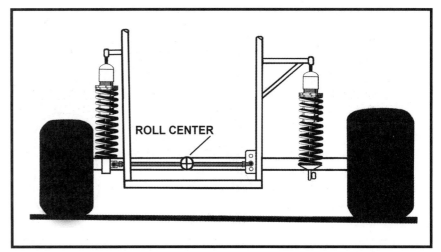

When suspension lateral movement is controlled with a Panhard bar, the roll center is located at the center line of the Panhard bar, both laterally and vertically.

The Roll Moment

A moment is a lever arm which rotates about a certain point. Relating to roll centers, a roll moment is the lever arm between the roll axis and the center of gravity height (CGH). The larger the roll moment, the greater the body roll during cornering. The greater the distance between the CGH and the roll axis, the greater the moment arm.

The moment arm length does not affect, by itself, the amount of weight transferred during cornering. It does, however, affect the car's resistance to body roll, and how it is distributed on the outside tires. The greater the moment arm length (or the greater the distance between the CGH and the roll axis), the less resistance there is to body roll. The smaller the moment arm, the greater the resistance to body roll. And, this affects the way in which the transferred weight is distributed to the tire contact patches.

As the roll center rises, the vehicle rolls less, and the weight transfer is applied to the outside tire contact patch in a less angular manner. Also, as the roll center rises, the spring rate required at the outside corners decreases because the roll stiffness rises with a shorter moment arm.

Let's examine the main advantages and disadvantages of both a high and low roll center and see why any design is a compromise. If there is a high roll center and thus very little body roll, softer springs can be used to control what body roll there is. The major disadvantage to a high roll center is that the lateral force during cornering, rather than exerting itself as body roll,

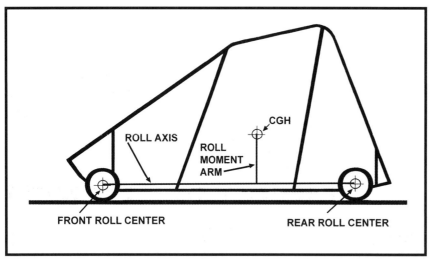

The roll moment arm extends between the roll axis and the CGH. It directly influences the amount of chassis roll during cornering. The larger the roll moment, the greater the chassis roll.

exerts itself in terms of lateral force at the outside tire contact patches, creating a shear force against the track surface. In other words, with a high roll center, weight is pushed sideways at the tire creating tire scrub.

With a low roll center, weight is transferred in a more circular motion, or in other words, it is picked up from the top of the inside wheel and forced down on top of the outside wheel. The advantage of a low roll center is less tire scrub, but the disadvantage is that there is more body roll to control.

Chapter 2

Suspension Systems

Front Suspension

All quarter midgets utilize a beam axle front suspension. A beam axle has both front wheels connected to the rigid axle. The beam axle is connected to the chassis with radius rods in the fore/aft direction, and with a Panhard bar in the lateral direction. This beam axle suspension is used because it is sturdy, and it is simple.

Caster

Caster in the front suspension provides directional stability. This influence is created with a line which is projected from the steering pivot axis down to the ground. This line strikes the ground in front of the tire contact patch when the caster is set in the positive position. A torque arm then exists between the projected steering axis pivot line and the center of the tire contact patch. This torque arm serves to force the wheel in a straight ahead direction. The greater the length of this torque arm (caused by greater amounts of positive caster), the greater the

Quarter midgets use a beam front axle with both front wheel connected to the rigid axle. It is connected to the chassis with radius rods in the fore/aft direction, and with a Panhard bar in the lateral direction.

steering effort required to turn the wheels away from their straight ahead direction. This keeps the car from wandering.

Positive caster, combined with the kingpin angle, creates another effect as well. When a car with positive caster turns left, the left front corner will rise and the right front corner will dip. As the car is steered right, the right front corner will

Suspension Systems

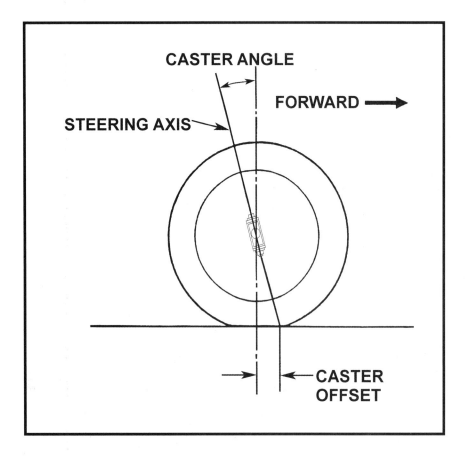

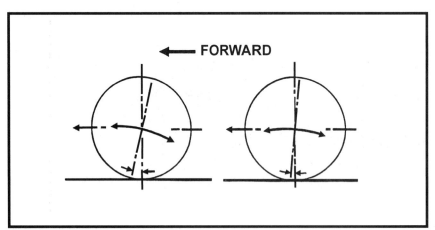

When a car with positive caster turns left, the left front corner will rise and the right front corner will dip. The amount depends on the amount of positive caster used combined with the kingpin inclination angle. The kingpin angle multiplies the effect. Notice how the curved path of the spindle changes with a large amount of positive caster (left) versus a smaller amount of caster (right).

rise and the left front will dip. The amount of these changes depends on the amount of positive caster used combined with the kingpin angle. The positive caster angle multiplies the effect of the kingpin angle and the associated corner lift and drop. The greater the positive caster, the more that the kingpin angle will change the corner height of the car as the wheel is steered. This effect is caused by the curved path that the spindle shaft follows as it is turned about the steering axis. This effect adds cross weight to the chassis during steering.

The more positive caster used, the more tracking feel there is for the driver. However, the more caster used, the harder the driver's steering effort becomes. If the driver experiences a lack of steering feel, or the car wanders too much down the straightaway, positive caster can be increased.

For younger, more inexperienced drivers, use a larger amount of positive caster – 5 to 6 degrees because it puts more feel in it for the driver. For more experienced drivers, use 4 to 5 degrees of caster. These settings are a starting ballpark. Work with what feels most comfortable for your driver. Ask for feedback, but your observations are also very important. Caster is a major element that affects the driver's comfort with the race car. Make sure you use this tuning element to make your driver comfortable. It adds a lot to his on-track confidence.

The caster is set at the right front by lengthening the lower radius rod and/or shortening

Suspension Systems

the upper radius rod. When setting caster, make equal turns on the upper and lower radius rods. For example, make one turn in on the upper and one turn out on the lower radius rod. Set the race-ready chassis on a flat, level surface, and place an inclinometer against the back of the bracket which holds the spindle. Adjust the radius rods to achieve the desired caster setting.

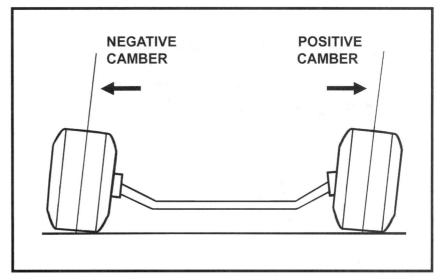

Camber

Camber is the inward or outward tilt of a tire at the top relative to vertical at the center of the tire. Zero camber is true vertical, negative camber is the tilt at the top of the tire toward the center of the car, and positive camber is the tilt of the top of the tire away from the center of the car.

Negative camber at the right front is used only to offset the effects of tire deformation during cornering. A beam axle keeps the camber angle constant between the tires and the race track. There is no change of camber during suspension travel with a beam axle as there is with an independent type of suspension.

The proper amount of camber maximizes the tire contact patch on the track. Optimum camber puts the entire tread on the track for even wear and grip. The normal range of camber is positive 1-degree at the left front, and negative 1 to negative 3 degrees at the right front. Heavier, faster classes usually need to use more static negative camber.

There are several different methods used to set camber. This spindle captures the top of the kingpin in a slot, and the screws at either end are turned in or out to set the desired amount of camber.

The caster angle has an effect on camber when the steering wheel is turned. As the car is steered the right front wheel gains a slight amount of negative camber, and the left front wheel gains positive camber. For both front tires, this adds camber in the proper direction for maximizing the

This type of adjuster allows both caster and camber to be set at the top of the spindle.

Notice how much negative camber is being used at the right front of this car. There is a gap under the outside edge of the tire, which means only the center and inside sections of the tread face are on the track. The tread wear pattern confirms this.

tread face of the tire during cornering.

The most efficient way of fine tuning the camber is by using tire temperatures. A tire pyrometer (or heat gun) is used to measure the inside, middle and outside temperatures of both front tires after the car has been run on the track. The heat distribution across the face of each tire will indicate if the tread face is working flat on the track surface, or if the tire is running more on the inside or outside edge of the tire. If the outside edge is hotter than the other two temperatures, that indicates the tire needs more negative camber. If the inside edge is hotter than the other two temperatures, that indicates the tire needs more positive camber. If the temperatures are virtually even across the tread surface, that indicates the tire is operating efficiently.

Kingpin Inclination

The kingpin inclination is a line drawn through the center of the spindle attachment bolt, extending to the ground. It is the inclined axis about which the spindle rotates to steer the wheel. The axis is inclined inward at the top toward the center of the car.

Kingpin inclination provides self alignment of the steering. This is because the kingpin inclination causes the chassis to rise when the wheels are steered, placing an additional weight loading on the spindle shaft.

The kingpin axis inclination has an effect over how the positive caster angle creates weight loading in the chassis during steering. Positive caster will cause the left front corner to rise up and add weight to that corner and the right rear when the car is steered to the left. When the car is steered to the right, positive caster causes the right front to rise up and add weight to that corner and the left rear. The kingpin axis angle multiplies this weight jacking effect. The greater the kingpin axis inclination, the greater the weight loading caused by positive caster.

On quarter midget cars, a 5-degree kingpin inclination is used for paved track cars, and a 10-degree kingpin inclination is used for dirt track cars. A greater amount of angle is used for dirt cars to increase the weight jacking effect when the car is steered.

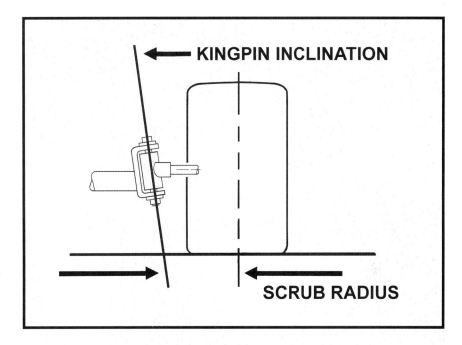

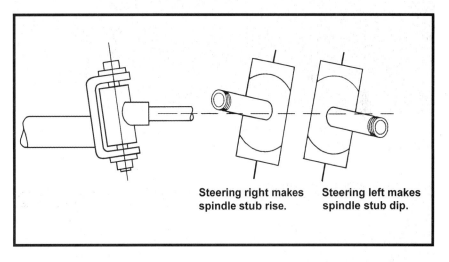

Steering right makes spindle stub rise. Steering left makes spindle stub dip.

Scrub Radius

The scrub radius is the distance from the kingpin axis line where it is projected to the ground, to the center of the adjoining tire contact patch. The scrub radius creates a leverage effect about the spindle. The amount of wheel offset and the kingpin inclination angle both have a bearing on the width of the scrub radius.

With a scrub radius, when you turn left, it lengthens the right side wheelbase, which tends to loosen the car. When you countersteer to the right, it shortens the right side wheelbase and it puts understeer in the car. This is a very stabilizing effect to the control of the car and it gives feedback to the driver.

When the right front wheel offset is moved outward one-half inch, it also increases the scrub radius one-half inch. The greater the scrub radius, the greater the arc of the spindle as it is steered. In combination with kingpin inclination and positive caster angle, the arc movement of the spindle serves to raise the front corner of the car up as it is steered to its extremes. The more scrub radius and the more kingpin inclination angle, the more it lifts the car.

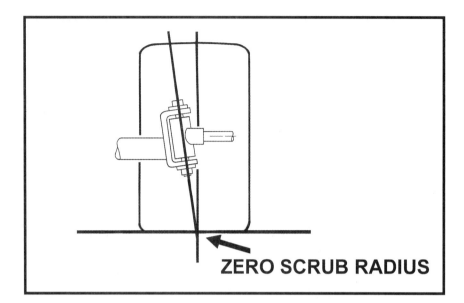

With a zero scrub design, the spindle must be tucked inside the wheel.

One important trade-off associated with having a larger scrub radius is the leverage it exerts on the steering wheel. When the right front wheel hits a hole in the race track, it creates feedback to the steering wheel. The larger the scrub radius, the greater the reaction is at the steering wheel.

A wider scrub radius also means that there is more tire tread face scrub as the tire is steered about the kingpin axis. The scrub radius offers a resistance as the tires are steered, which gives steering feel and feedback to the driver.

Some chassis designers advocate a zero scrub radius or limited scrub radius design. This makes the tire centerline and kingpin axis intersect, providing no scrub radius, or a very minimal scrub radius, as the tire pivots at its centerline. This is accomplished by tucking the spindle closer into the wheel.

The problem with zero scrub design is that there is no scrub radius, and thus no driver feel or feedback. The scrub radius width provides the driver feedback with steering feel. If the front end pushes or the front tires hit a slick spot, the driver will feel a change in the steering. The wider the scrub radius, the more feel and feedback. The smaller the scrub radius, the lighter the steering feel. The drawback to a wider scrub radius is more tire surface scrub during steering.

Ackerman Steering

Ackerman steering geometry is created when the inside front wheel is steered at a greater angle than the outside front in order to eliminate tire scrub at the inside wheel during cornering.

The direction the left front tire steers without tire scrub is very important. It must steer at a sharper arc than the right front because it is traveling on a shorter radius. The left front simply steers more and helps point and stabilize the car. Increased Ackerman helps decrease — or eliminate — cornering understeer at turn entry and through the middle of the corner. Ackerman steers the left front wheel more and thus helps to point the car.

Ackerman steering actually is dynamic front toe-out gain. It only creates toe-out as the front wheels are steered. Large amounts of static toe-out are not beneficial to the car because it creates excessive drag on the straightaways and causes darty steering response. Ackerman steering will only create toe-out when the car is steered. Ackerman is gained directly in proportion to how much the steering wheel is turned.

The amount of Ackerman steering a car has is measured as the difference in steering angle between the right front and the left front wheels. Because on oval tracks the right front is the controlling tire, the amount of Ackerman steering is quoted as the amount of steering angle gain at the left front over the right front.

In general, the left front has to turn about 15 percent more than the right front. This means that if the right front tire turns 10 degrees, the left front is going to have to turn 11.5 degrees to minimize tire scrub. Therefore, a car requires 2 to 3 degrees of toe-out gain at the left front in 18 degrees of right front steering. In other words, if you turn the right front wheel 18 degrees to the left, the left front should show 20

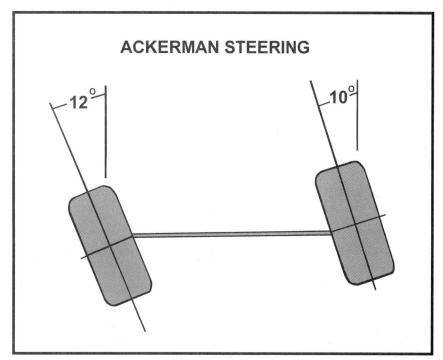

Ackerman steering geometry is created when the inside front wheel is steered in a sharper arc than the outside front during cornering in order to eliminate tire scrub at the inside wheel.

Ackerman steering will only create toe-out when the car is steered. The toe-out gain is at the left front. Ackerman is gained directly in proportion to how much the steering wheel is turned.

degrees of steering angle. The tighter the turn radius, the more the left front has to turn. The wider the turns – like on a wide sweeping track – the less Ackerman gain required at the left front.

Ackerman steering on a quarter midget is most commonly achieved through the use of skewed steering shaft ears. The ears are attached to

Ackerman steering on a quarter midget is commonly achieved through the use of skewed steering shaft ears. The ears are attached to the steering shaft at 1 o'clock and 11 o'clock.

Another method of creating Ackerman steer is to use a shorter spindle steering arm on the left side than on the right side. The problem with using a shorter left front arm is that toe-in is created when the front wheels are steered to the right.

the steering shaft at 1 o'clock and 11 o'clock. Looking at the steering shaft ears from the driver's seat, the ear which attaches the right front steering rod is attached to the 11 o'clock ear, and the ear which attaches the left front steering rod is attached to the 1 o'clock ear. This arrangement pushes on the left front spindle arm more than it pulls on the right front spindle arm.

Using Ackerman in the chassis not only is helpful for turning the car into a corner, but also for countersteering when a car is loose. When a car is countersteered, Ackerman will steer the right front tire outward more. This will help steer the nose of a car out of a spin. However, if there is no Ackerman or the front wheels toe in when countersteered to the right, the front end of the car will be pinned and it will have a much greater tendency to spin out.

Anti-Roll Bar

Some quarter midget chassis manufacturers design their cars to use an anti-roll bar in the front suspension. An anti-roll bar is a torsion bar connected with arms to the front axle on each side. The reason for using the bar is to allow softer front spring rates to be used. The anti-roll bar only comes into play when the chassis rolls as the car corners. On straightaways, the bar does nothing. During cornering, both of the front suspension springs and the anti-roll bar are used to control, or resist, chassis roll. The total spring rate of the two suspension springs and the anti-roll bar added together is called the roll rate.

Race cars also have a "ride rate." This is the combination of the suspension springs that control, or resist, the straight up and down movement of the chassis over the suspension.

On paved track cars, more roll rate is required than ride rate. This is because paved track cars experience higher G force loadings in the corners than on a dirt track. The tires stick well and have little slip. Consequently, there is more chassis roll to be resisted. But the ride rate cannot be called upon to handle the roll rate as well. The spring rates on paved track cars should be as light as practical. So, a "roll helper spring," or anti-roll bar, is added to supplement the rate of the suspension springs to add roll rate.

Rear Suspension

All quarter midgets utilize a live rear axle. A live rear axle has both rear wheels connected to one rigid axle. It is called a "live axle" because one-wheel bump affects the opposite wheel as well. The rear axle is connected to the chassis with radius rods in the fore/aft direction, and with a Panhard bar in the lateral

Some quarter midget chassis manufacturers design their cars to use an anti-roll bar in the front suspension. Using the bar allows softer front spring rates to be used.

All quarter midgets utilize a live rear axle. It is called a "live axle" because one-wheel bump affects the opposite wheel as well. The rear axle is connected to the chassis with radius rods in the fore/aft direction, and with a Panhard bar in the lateral direction.

(Above and below) The four link rods move the axle in a curved path as the axle moves in bump and rebound. This will pull the axle forward during suspension travel.

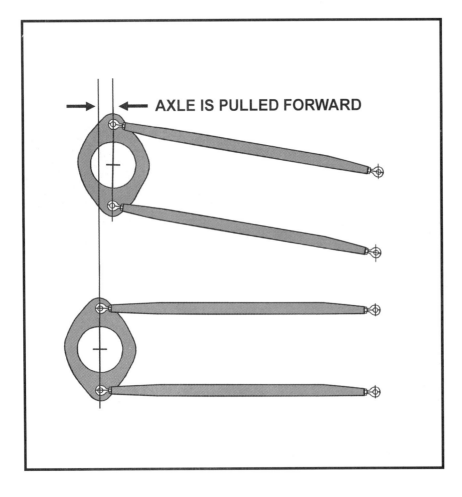

direction.

Four Link Suspension

The four link rear suspension uses two equal length parallel radius rod links on each side that go forward from the birdcage to the chassis.

Four link rods move the axle in a curved path as chassis roll occurs. On the right side, the axle and bearing carrier move up as the right rear suspension moves in bump travel. The curved path of the axle centerline travel pulls the right side of the axle forward.

On the left side, the axle and birdcage move downward in compression travel as the chassis rolls to the right. The curved path of the axle centerline travel also pulls the left side of the axle forward. But on most quarter midget suspension designs, the left side radius rods are shorter than the right side rods. This means that the left side birdcage and axle centerline moves in a sharper arc than the right side. So, the left side of the axle is pulled forward more than the right side of the axle.

This creates a slight amount of roll steer as the car corners. When the rear axle is skewed with the right side pushed back and the left side pulled forward, the axle is steering the rear of the car toward the outside of the corner. This is called roll oversteer. The amount of roll steer present is dependent on the amount of

Suspension Systems

chassis roll the car experiences.

Unlocked/Locked Left Rear

All classes except Light B, Light AA, Heavy AA, and half use an unlocked rear axle (and in some cases these cars use an unlocked left rear). "Unlocked" means the left rear wheel is not locked together with the right rear wheel through the axle. The left rear turns freely independent of the right rear. A locked rear axle means the left rear and right rear wheels are locked together through the axle.

The reason that most classes use an unlocked left rear is that quarter midgets are very low-powered race cars. If the rear axle was totally locked up – the left rear and right rear tires driving together – it would produce way too much tire drag, and bog down the engine as it accelerates through mid-corner and corner exit. Unlocking the left rear frees up the chassis.

By QMA rules, the left rear wheel can be locked up in any class.

On dirt (or slick pavement), locking the left rear helps to stop an oversteer slide because the left rear helps to drive the car forward.

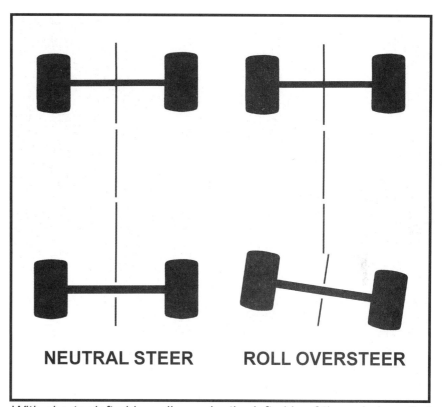

With shorter left side radius rods, the left side of the axle is pulled forward more than the right side of the axle. This creates a slight amount of roll oversteer as the car corners. When the rear axle is skewed with the right side pushed back and the left side pulled forward, the axle slightly steers the rear of the car toward the outside of the corner.

The left rear hub is locked to the axle by placing a key in the keyway slot between the axle and the hub.

Suspension Systems

The left rear coil-over can be mounted in front of (below) or behind (above) the rear axle. When it is mounted behind the axle, the car has a longer spring base and a softer left rear spring rate can be used.

Left Rear Ratchet Hub

A ratchet hub is sometimes used at the left rear instead of a locked or unlocked hub. The ratchet hub disengages when the driver lets off the gas, so it functions just like an unlocked rear. When the driver applies power, it engages and fully locks the left rear hub to the axle.

Left Rear Coil-Over Location

The left rear coil-over can be found mounted either in front of or behind the axle. The mounting position depends on the track surface. The forward mounting position is used on smoother tracks. On rougher, bumpier track, it is best to mount of the coil-over behind the axle. Mounting the coil-over behind the axle creates a longer spring base, which means that a softer spring rate can be used. When the coil-over is mounted forward of the axle, it is supporting much more weight, so the spring rate must be stiffer.

Roll Centers

Roll centers (front and rear) are important design parameters in the race car. Together with the center of gravity height, these parameters influence everything the car does, the way the car behaves, and the effects of suspension changes you make to the chassis. Roll center heights can be changed – both front and rear – to adjust the chassis for oversteer and understeer.

The effect of centrifugal force during cornering acts on the total car mass through its center of gravity. This force is experienced as "lean" or "sway," but is most commonly

Suspension Systems

known in vehicle dynamics as "roll." The effect of this centrifugal force on the center of gravity is dependent on speed (velocity). The higher the center of gravity, the lower the speed at which a car will reach its maximum roll.

When the chassis rolls, just like anything in a circular motion, it must have a central point about which it rotates, or rolls. In a car this is known as the roll center.

In any car there are two roll centers – front and rear – and they are independent of each other. But when a car is cornering, the front and rear suspensions are connected through the chassis and they both experience roll at the same time. So an imaginary line called a roll axis connects the front and rear roll centers.

The center of gravity height of the car is connected to the roll axis through a roll moment arm. A moment is a lever arm which rotates about a certain point. Relating to roll centers, a roll moment is the lever arm between the roll axis and the CGH. The larger the roll moment, the greater the body roll during cornering. This is because of a force called overturning moment, which is the cornering load transfer being accelerated through the roll moment. The greater the distance between the CGH and the roll axis, the greater the moment arm, and the greater the overturning

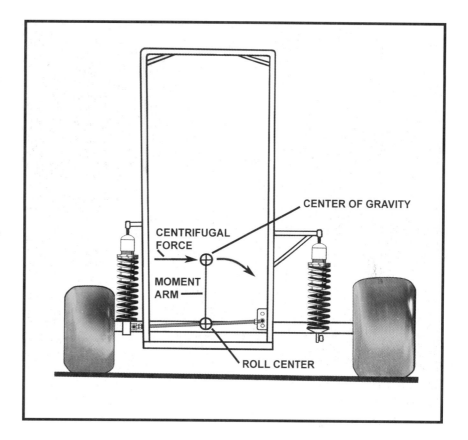

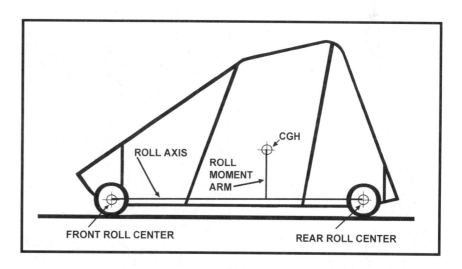

moment which transfers weight from the inside to the outside during cornering. So, when the height of the roll center is changed, it affects the car's resistance to chassis roll and how the transferred weight is distributed on the outside tires.

The roll center on a beam axle suspension is established by the linkage which locates the axle against the body's lateral (sideways) movement.

The best type of chassis mount for a Panhard bar is a sliding clamp mount (left). This allows the bar to be adjusted in small increments for fine tuning. A change as small as 1/8-inch can make a difference in handling. If the Panhard bar mounts to a bracket with a series of holes (right), the amount of adjustment is dictated by the spacing between the holes (usually 3/8-inch). Remember that this type of adjustment is only a fine-tuning adjustment, so don't get carried away with Panhard bar adjustments.

The front roll center is located at the center of the Panhard bar, both vertically and horizontally. At the rear of the car, the rear roll center is located at the center of the Panhard bar.

This is the Panhard bar at the front and rear suspension.

Front Roll Center

On the front axle, the Panhard bar attaches to the axle on the left side and to the chassis on the right side. The center point of this Panhard bar is the front roll center height of the car. The roll center is the point about which the sprung weight of the vehicle rolls during cornering. The higher the roll center, the less chassis roll, because the roll center is closer to the center of gravity height. The lower the roll center, the more chassis roll.

If the front Panhard bar height mount is adjustable, lowering the front roll center will make the front of the car stick better by creating more overturning moment onto the right front tire. That creates more downward loading onto the right front. If the front roll center height is raised, it will loosen the front end and tend toward giving the car a push.

An initial starting point for the front Panhard bar mounting height is the frame end of the bar 0.5-inch higher than the axle centerline. This makes the Panhard bar parallel with the front axle as the chassis rolls during cornering. Start with this setting unless your chassis manufacturer has another starting specification. The front Panhard bar can be moved up or down within a certain range to adjust for handling conditions.

Rear Roll Center

On the rear axle, a Panhard bar is the linkage commonly used as the lateral locating device between the chassis and the axle. The bar mounts to the birdcage on the left side and to the chassis on the right side. Because the chassis rolls to the right during left-hard cornering, this means the rear roll center is lowered during cornering. A lower roll center helps to tighten up the rear of the car during cornering.

A Panhard bar provides control over right rear tire grip during cornering by raising or lowering the rear roll center height. Lowering the Panhard bar (and thus the rear roll center) helps to tighten up the chassis during cornering. Raising the Panhard bar (and thus the rear roll center) helps to loosen the chassis during cornering. Many chassis manufacturers make the Panhard bar chassis mount a clamp bracket type of mount so it is easy to move the bar up and down and fine-tune the adjustment.

Spring Rates

The spring rate of a spring is the comparative rating of its resistance to a load placed on

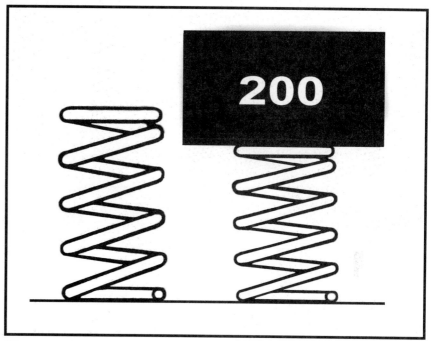

The rate of a spring is the amount of force or weight needed to deflect a spring a given amount. For example, if 200 pounds were placed on a spring and it compressed one inch, the rate is 200 pounds per inch.

it. This measurement is expressed in terms of pounds per inch. In other words, if 100 pounds were placed on a spring, and it compressed one inch, the rate of the spring is 100 pounds per inch (the load divided by inches of compression). If 100 pounds were placed on the spring and it compressed two inches, the rate would be 50 pounds per inch.

The example above defines a linear spring rate. That means each equal increment of weight placed on the spring would compress the spring the same distance as the first increment. In other words, in the case of the 100 #/" spring, 100 pounds placed on it would compress it one inch, so 200 pounds placed on it would compress it two inches. Only linear rate springs are used in race cars. If a progressive rate spring (one that gets stiffer the more that it is loaded) was used, its spring rate would vary as the car experienced more or less chassis roll during cornering.

Softer spring rates allow the tires to maintain a better compliance with the race track. Better tire compliance with the track surface means the tires ride over bumps on the surface easier. There is more energy stored in a stiffer spring than a softer spring, so a stiffer spring produces more rebound after being compressed. With less stored energy, the softer spring allows the tire to negoti-

Suspension Systems

A coil-over is a suspension unit that uses a shock absorber with a coil spring mounted over it. The coil spring is contained over the shock with an adjustable collar.

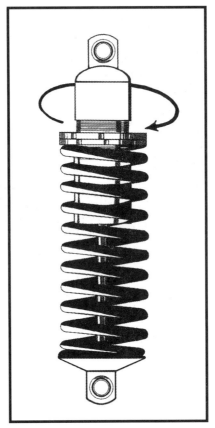

Screwing the adjustable collar up or down on the shock body will change ride height and weight loading at that corner of the car. To add weight (or "jack weight in") or raise the corner height, turn the adjusting collar down (clockwise).

ate track irregularities more easily, providing less tire bounce.

Spring rates also determine how much weight load transfer is resisted at each wheel during cornering, braking and accelerating. Spring rates can be adjusted stiffer or softer to loosen or tighten the car as needed to achieve the desired handling balance. Spring rates can help change the handling balance between neutral, loose or tight.

Once the spring rate balance is found, the handling can be fine-tuned at the track. See the *"Track Tuning & Adjustment"* chapter for more details.

Coil-Overs

A coil-over is a suspension unit that uses a shock absorber with a coil spring mounted over it. The coil spring is contained over the shock with an adjustable collar. Screwing this adjustable collar up or down on the shock body will change ride height and weight loading at that corner of the car. To add weight (or "jack weight in") or raise the corner height, turn the adjusting collar down (clockwise). To take weight out, or lower the corner height, turn the collar up (counter clockwise).

Sprung Vs. Unsprung Weight

Unsprung weight is the weight of suspension pieces which are not supported by the car's springs. Examples of these are tires, wheels, beam axles, hubs, and the brake.

Suspension pieces which are hinged from the chassis contribute only one-half of their weight to unsprung weight. Because they are attached to the chassis, the other half of their weight is considered sprung weight. Components in this category include radius rods, coil-overs and steering tie rods.

Sprung weight is the weight of the chassis which is supported by the springs.

How do you determine what is sprung and what is unsprung weight on a car? A very quick and graphic explanation is to lift the race car up by the roll cage. What goes up with the chassis is sprung weight. What drops down is unsprung weight. And what keeps the unsprung components attached to the sprung is half unsprung and half sprung weight.

Controlling, or limiting, unsprung weight is probably ten times more important than reducing sprung weight. Why is unsprung weight so much more important? Think of a racing tire mounted on a 300-pound concrete wheel, and think of a small lightweight compact car tire on a light rim. Both tires go rolling down a road and hit a large bump. The big tire on the concrete wheel (the heavier mass) will go higher and accelerate faster than the lighter tire and wheel, and it would be off the ground for a longer period of time. Now think of being in front of these two tires right after they hit the bump, and trying to catch them. The lightweight wheel would probably be a fairly easy catch, but the heavy and rapidly accelerating racing tire/concrete wheel would probably be impossible to control.

And so it is with unsprung weight. The lightweight tire and wheel relates to light unsprung weight; the racing tire and concrete wheel relate to heavy unsprung weight. The springs and shock absorbers are used to control the unsprung weight. The heavier the unsprung weight, the heavier the spring rate and stiffer shock absorber rate required to slow it down. The shock absorber slows its movement while the spring

Using lighter weight suspension pieces saves unsprung weight. A 34-inch titanium axle weighs only 2.5 pounds, whereas a 34-inch steel axle weighs 8 pounds.

pushes it back down. But the suspension will be so stiff that it won't let the car transfer sufficient weight. And, this will also heat up shocks and cause them to fade.

Heavier unsprung weight will cause a particular wheel to be off the ground longer. The point here is that no matter what the overall weight of the vehicle is, the unsprung weight should be kept to an absolute minimum. The rougher the track surface is, the more critical it is to control unsprung weight.

Lightweight suspension components help to decrease unsprung weight. With less unsprung weight, forward traction is improved, the tires wear less and the car has a better exit speed because the tires are staying on the track better.

An example of lighter suspension pieces would be using an aluminum rear axle in place of a steel axle, or a titanium axle in place of an aluminum axle. Typically, a 34-inch steel axle weighs 8 pounds, a 34-inch aluminum axle weighs 5 pounds, and a 34-inch titanium axle weighs 2.5 pounds.

Weight Adjustment

Corner weights of the race car are adjusted using the coil-over adjusters. A very important principle to understand is that when weight is adjusted at one corner, it does the same thing to the diagonally opposite corner of the car. And, it affects the other two corners of the car as well.

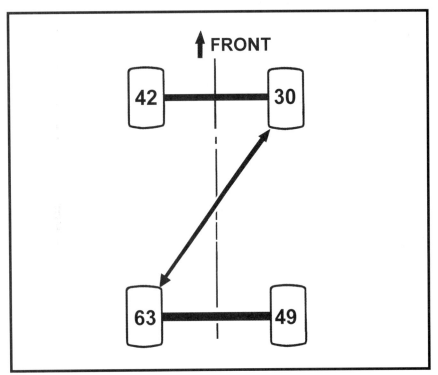

If the right front coil-over adjuster is used to add weight to the right front corner, it will also add weight to the diagonally opposite corner, which is the left rear. And this adjustment also affects the left front and right rear corners by taking weight off of these two corners.

For example, if you use the right front adjuster to add weight to the right front corner, it will also add weight to the diagonally opposite corner, which is the left rear. And this adjustment also affects the left front and right rear corners by taking weight off of these two corners.

Using the coil-over weight adjusters will change corner weights, but these adjustments will not change the rear weight percentage or left side percentage. That is because adding corner weight at the right front adds weight to the left rear and subtracts an equal amount from the left front and right rear corners. To change left-to-right and front-to-rear weight percentages, the weight mass placement in the chassis must be moved.

Chapter 3

Shock Absorbers

The Purpose of Shocks

Shock absorbers – what are they? First of all, they are not shock absorbers! (The United States is the only place that calls them that. The rest of the educated world calls them "dampers.") Springs absorb shock. Shocks really dampen the kinetic energy that is stored in the springs. The energy is created by vertical movements of the wheels relative to the chassis. The kinetic (moving) energy is converted by the shock absorber to heat energy by creating resistance to movement within the shock.

Shock absorbers are velocity sensitive, heat dissipating, hydraulic devices. The shock absorber offers resistance to movement of the suspension by forcing hydraulic oil through a series of valves and openings inside the shock.

Shocks are intended to control the vertical oscillations of the suspension caused by bump and dynamic load transfer. Without shock absorbers, when a bump or load transfer is encountered, the vehicle would continue to oscillate or bounce.

Double Tube, Low Gas Pressure Shocks

There are two different type of shock absorbers commonly used in race cars – the high pressure gas shock (also called the monotube or single tube shock), and the low gas pressure shock (also called twin tube or double tube shock).

Double tube shocks utilize two tubes – an inner cylinder where the piston operates against the shock fluid, and the outer shock body. The space between the two cylinders is utilized as a reservoir.

A shock cannot be fully filled with fluid. This is because the fluid has to have some place to go as the shaft moves inward in compression which subtracts from area inside of the cylinder. The fluid is non-compressible, so it has to have a chamber to travel to. In the double tube shock, that space is between the inner tube and outer tube.

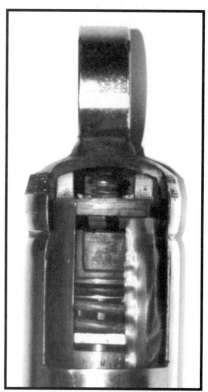

This cutaway of a double tube shock shows the outer tube, the inner tube and the piston and shaft inside the inner tube. The space between the outer shock tube and the inner tube is the reservoir area. Visible at the right between the two tubes is the gas-filled plastic bag.

If that space was filled with just plain air, a big problem would occur: aeration. Aeration occurs when air

The piston and shaft assembly from a double tube shock. The heavy spring is a compression bypass spring valve. The holes in the piston are outlets for bleed valves and high speed jets. The milled slots in the piston control low speed bleed.

The foot valve at the bottom of the inner cylinder of a double tube shock contains two spring loaded valves. The one on the left is a blow-off valve that controls compression damping. The one on the right is a directional control valve. It pulls open under rebound to fill the cavity under the piston. The milled slots function as low speed bleeds to allow fluid transfer from and to the reservoir.

mixes with the oil and bubbles form in it. Aeration makes the shock fluid compressible and changes its viscosity, and thus lowers the resistance of the flow through the valves and orifices. This takes away most of the control characteristics of the shock.

To prevent oil aeration, the reservoir space between the inner and outer tubes is filled with a gas-filled plastic bag. The bag contains nitrogen, which is compressible. The gas bag allows for fluid volume changes as the shaft moves in compression and rebound. The gas bag volume changes during shaft movement. The plastic bag totally contains the nitrogen and prevents it from mixing with the shock fluid.

Monotube High Gas Pressure Shocks

Monotube high gas pressure shocks are designed differently from twin tube shocks. They have a single tube, they use deflective discs for valving, and they have a high pressure nitrogen gas chamber which is used to resist oil cavitation (foaming). The monotube functions as the inner bore for the piston, as well as the outer shell of the shock body. However, when the shock body becomes dented, the piston movement is hampered, thus resulting in shock failure. But, a benefit of the monotube design is that excessive heat from the oil transfers to the outer surface of the shock body more efficiently than in a twin tube shock, so heat is dissipated more efficiently.

A positive aspect of the monotube is that the design allows a larger diameter piston to be used since the bore is larger. This design also requires a second floating piston that is located between the shock oil and the pressurized gas. The gas pressure is what creates enough force to control oil cavitation (foaming). When pressure on the shock oil gets low enough, bubbles will form in the oil as the piston moves through it. This diminishes the resistance force that the shock provides.

The volume inside the monotube shock is variable because the gas is compressible. When the shock is compressed, the shaft is pushed into the shock body which increases internal volume. The shock tube is completely sealed and the oil is non-compressible. So something has to give to accommodate the increased internal volume. What happens is that the gas is compressed.

When the shock is extended (rebound), the volume in the oil chamber decreases as the shaft moves out. But pressure still has to be maintained in the internal chamber so that bubbles do not form in the oil as the main piston is pulled back through the oil. That means there has to be enough pressure on the gas side of the floating piston to maintain sufficient pressure against the oil to prevent cavitation as the gas chamber volume increases.

Using more or less internal gas pressure makes a monotube shock react faster or slower. A higher gas pressure makes a shock react faster. Less internal gas pressure makes a shock react slower.

Shock Absorbers and Handling

Shock absorbers are very important to a car's handling. Shock absorbers control the rate of weight transfer during cornering, they control spring movement, and they control suspension movement over bumps and surface irregularities. Being able to control the chassis with the proper shock absorbers is a key element to proper handling. Shocks can be used to help control handling problems or to induce desirable handling characteristics.

Shocks have no effect on the *amount* of weight that is transferred dynamically during braking, acceleration and cornering. They can, however, affect the transient response in the pitch and roll axis. The amount of weight transferred is dependent on the center of gravity, roll axis, and roll rates.

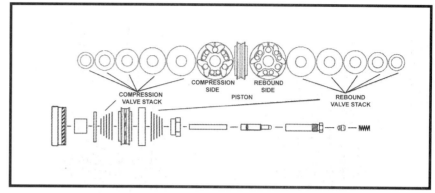

Monotube high gas pressure shocks use deflective discs for valving. The discs are spring steel shims which are attached to the piston. Fluid flowing through the metering orifices in the piston must deflect the shim stack, which uncovers other orifices in the piston to alter the pressure build-up. A shim stack resides on either side of the piston, so one stack is deflected during compression, and the other one is deflected during rebound. The damping characteristics are determined by a given amount of resistance at a given piston velocity. These characteristics are tailored by the shape, diameter and thickness of the steel shims.

*Shocks have no effect on the amount of weight that is transferred dynamically during braking, acceleration and cornering. They can, however, affect how **quickly** the weight is transferred. A softer shock allows weight to transfer quicker than a stiffer shock.*

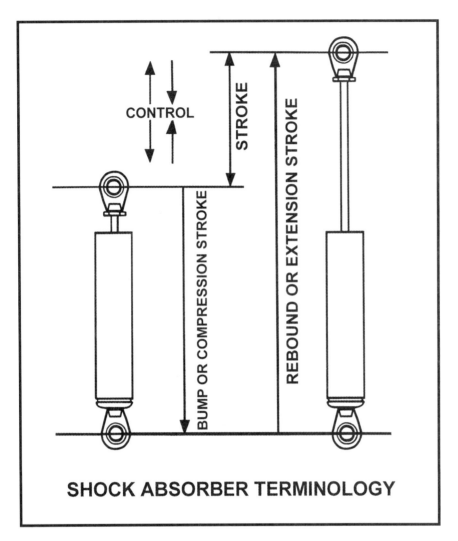

SHOCK ABSORBER TERMINOLOGY

Where the weight is transferred is dependent on spring rates. How quickly the weight is transferred is controlled by the shock absorbers. This may only be for an instant, but shocks play an important part in handling response during that instant. It is the low speed damping forces of a shock that most influence the chassis as it rolls or pitches (roll is corner entry/exit, pitching is acceleration/deceleration).

How Shocks Influence Handling

Inside shocks, there are a series of valves and orifices, or deflective discs. When the shaft and piston assembly move, they force fluid through these. This creates a resistance to movement. Shocks will produce a resistance force which is proportional to the speed of the shaft movement. When compressed slowly, a shock generates less resistance force than when it is moved faster. This principle is used to create different levels, or stages, of resistance force. This staged valving is necessary because the shock resistance required to control the race car suspension when it goes over a severe bump (referred to as high speed control) is much greater than the resistance needed to control chassis roll or the suspension movement caused by small bumps (referred to as low or medium speed control). Shock resistance at low, medium, and high piston speeds must be matched to the needs of the race car.

In the simplest terms, racing shocks perform two functions:

1. When bumps and ruts are encountered, shocks keep the chassis settled and the tires in compliance with the race track. Without shocks, the chassis would pitch, roll, and bounce violently whenever the race car encountered bumps and ruts. The tires could lose contact with the track surface.

2. Shocks help control the rate of chassis roll and pitch caused by dynamic weight transfer. Whenever a race car accelerates, decelerates, or corners, the chassis will pitch or roll (due to weight transfer). Without shocks, body roll and pitch would be violent and the chassis would not be stable.

Shock control at low piston speeds affects how the race car handles through the corners, while medium and high speed control affects how the race car handles whenever it encounters bumps and ruts.

Rebound control is a shock's resistance to extending and is specified at a given piston speed. The amount of rebound control developed by a shock will determine, generally, how quickly a tire is unloaded during dynamic weight transfer, and how quickly the suspension "rebounds" or returns to its original position, after the spring has been compressed.

Compression, or bump control, is a shock's resistance to compression and is specified at a given piston speed. Compression control will determine, generally, how quickly a tire is loaded during dynamic weight transfer and how the suspension will react whenever a bump is initially contacted.

The stiffness of the shock absorbers used on a race car has a profound effect on the rate at which weight transfer affects the loads on the tires. The "rate" of weight transfer means how quickly or slowly a particular shock valving allows weight to transfer. Because of this, shocks are a very important factor when it comes to handling. Basically, soft shocks allow weight transfer to affect tire loadings more quickly than stiff shocks.

There really is no mystery to shock function and tuning. However, there are complexities and qualities that need to be considered when choosing shocks for a specific application. By keeping this basic information in mind you should be able to install the correct shocks for each situation when troubleshooting handling problems. This should also enable you to have the confidence to make changes with fairly good expectations of results.

Above all, remember that chassis tuning is a compromise and shocks, though a very important part of the setup, are still only a part. Keep the following in mind for proper chassis tuning:

1. As the piston speed of a shock increases, the shock gets stiffer.

2. Large bumps hit at high speeds cause the highest piston velocities, and the highest shock resistance, to occur.

3. The low speed resistance of a shock absorber controls the rate of body roll and pitch, and also how quickly a tire is loaded and unloaded during dynamic weight transfer.

4. Generally, soft shocks will cause a tire to become loaded or unloaded (due to dynamic weight transfer) more quickly than stiff shocks.

Shock Damping Stiffness

The term damping, also called dampening, means to control the forces of suspension compression and rebound travel. Compression is the travel direction of the

The low speed resistance of a shock absorber controls the rate of body roll and pitch, and also how quickly a tire is loaded and unloaded during dynamic weight transfer.

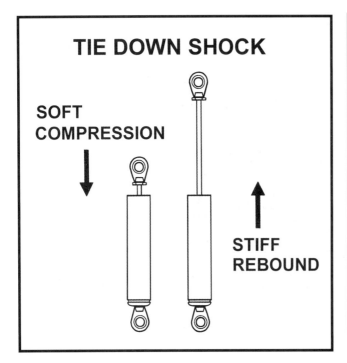
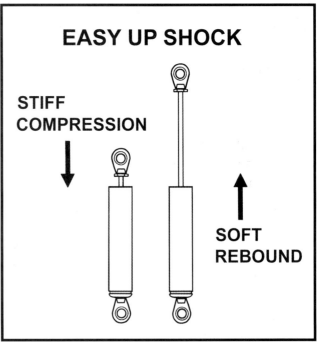

shock getting shorter – the shaft travels into the shock body. Rebound is the opposite. Rebound travel is the extension of the shock absorber.

The compression and rebound movements of the suspension are dampened, or controlled, by the shocks. Shocks provide this damping control through hydraulic resistance. The movement of the shock absorber forces oil through a series of valves and orifices in the body. These valves and orifices can be designed to provide any amount of resistance to movement.

Shock absorbers are engineered to provide specific measurable amounts of resistance, or damping control, in rebound and compression. This is done by equipping a shock with different internal valves (thus the term shock valving). In general terms, the damping force ranges from soft (meaning very little resistance or control) to stiff (meaning heavy resistance or control). The amount of damping control a shock provides is designated by damping stiffness codes.

Damping Stiffness Codes

Manufacturers of shock absorbers place a number on them to indicate the relative stiffness of the shock's damping force. This is called the valve code.

Most quarter midget shocks are designated as a #3 (softer) or a #5 (stiffer) valving code. They are called 50/50 shocks, which means they have the same amount of damping stiffness in compression and rebound. In most cases, however, these shocks will have a somewhat higher amount of rebound damping stiffness than compression stiffness.

The "take apart" type of monotube shock can be valved for any amount of rebound and compression damping desired. These shocks use deflective shims, or valve stacks, that are located above and below the main piston. By changing the thickness of individual shims, the damping forces are altered.

Split Valving Shocks

There are conditions when you want to use two different shocks rolled into one. For example, you may want to keep a particular corner of a car from transferring a lot of weight when it rises up, which would require a stiff rebound valving shock, but you don't want to have a stiff shock for compression or bump travel at that corner. In that case, a split-valving shock can be used. A split valving shock has one rate in compression, and another rate in rebound.

A shock that is stiff on rebound valving and softer on compression valving is called a "tie down" shock. It keeps the corner of the car where it is attached tied down, making it more difficult for the chassis to rise up at that corner. Tie-down shocks are many times used on the left rear corner of heavier, faster cars such as "B" and "AA." Because of the amount of weight that these cars transfer at corner entry, the right front tire can become overloaded at corner entry. A tie-down shock at the left rear slows the weight transfer to the right front. A quarter midget tie-down shock is typically a #2 valving in compression and a #5 valving in rebound.

The opposite of a tie-down shock is an "easy up" shock. It has a rebound damping that is soft, and a compression that is a stiffer valving. A quarter midget easy-up shock is typically a #5 valving in compression and a #2 valving in rebound. The easy-up shock is used on a corner of the car when you want to have weight transfer very quickly to the opposite end of the car. For example, on a slippery paved track or a dry slick dirt track, on acceleration you want to have weight transfer from the front to the rear wheels very quickly. To accomplish this, the easy-up (soft rebound) shock is used at both front corners. On corner entry, the stiffer compression valving slows weight transfer onto the right front, so the right front tire is not overloaded and result in a push.

An important reminder: the basic shock setup should always start with 50/50 shocks, and not split valving shocks. When you put a split valve shock on a car before you sort out the springs, that shock may cover something up. You will spend time trying out different shock combinations, when the real problem may be with spring rates. Always start with 50/50 shocks and get the chassis sorted out with the correct springs and weight distribution. Then use split valve shocks to fine tune the chassis.

How Shocks Influence Handling

The rate of weight transfer during cornering affects how the car reacts to pitch and roll. Different combinations of shock absorber compression and rebound damping will influence handling through the three different segments of

At corner entry, the car is transferring weight primarily from the left rear to the right front. The rate of weight transfer can be influenced by shock damping. If the car is transferring too much weight too quickly, stiffen the right front compression and the left rear rebound damping.

cornering. Those segments are: 1) corner entry, 2) mid-corner, and 3) corner exit. Each segment of the corner will be affected by the amount of shock rebound and compression damping at each corner of the car.

Segment 1 – entering a corner. This is the first third of a corner. The car is transferring weight primarily from the left rear to the right front. This is influenced by rebound damping at the left rear and compression damping at the right front. If the car is transferring too much weight too quickly, stiffen the right front compression and the left rear rebound. Stiffening the shock damping means increasing the valving code to a higher number.

Segment 2 – mid-corner. At this point, the left front corner of the car is rising up and the right rear corner is being compressed. Control over these movements is influenced by rebound damping at the left front and compression damping at the right rear. If the car is too tight during this cornering segment, increase rebound damping at the left front and compression damping at the right rear. If the car is loose, use less rebound at the left front and less compression at the right rear.

Segment 3 – corner exit. The desire here is to get weight to transfer to the rear of the car to drive the car off the corner. Whether the car has a locked or unlocked rear axle, more left rear weight tightens up the car under acceleration at corner exit. If the track is slick and the car is loose, you want weight to transfer from the right front to the left rear. A softer left rear compression and softer right front rebound damping will quicken the weight transfer to the left rear tire during acceleration at corner exit.

Chassis Tuning With Shocks

The following are basic handling adjustment tips for tuning a chassis with shock absorbers. As you read them, think about the basic concept of what the shock is doing to the chassis. This way you will be able to make decisions at the track to cure handling problems that your car may encounter.

As you read through these shock absorber chassis tuning guidelines, try to picture in your mind how the car is reacting at each position on the track, and how the prescribed changes can affect the chassis. Visualize how the weight is transferring at each corner and at each position on the track, and how shocks can restrict or accelerate transfer at a particular track position. Once you understand how the weight transfer and transitional control process works, you will have a good understanding of how to fine-tune your chassis with shocks.

Loose At Corner Entry

Use stiffer left front compression so the chassis doesn't drop suddenly on the left front and make the car turn in early. Less rebound on the left rear will also help tighten the chassis at corner entry. When the right rear compression damping is decreased, it will load the right rear spring faster, which will help to stick the right rear tire at corner entry.

Tight At Corner Entry

Increase rebound on the left rear. This keeps weight on the left rear longer during corner entry, preventing a quick transfer onto the right front corner. This will loosen the chassis.

If the push is more severe, also use a tie down shock at the left front. This will allow the left front to compress more easily and keep the corner tied down. That allows the car to drop down onto that corner quicker and makes the car turn in easier.

Loose At Mid Corner

1) Use softer left rear rebound. This helps to continue loading the right rear from corner entry.

2) Use softer right rear compression. This allows weight transfer to roll over onto the right rear to stick that tire. A softer left front rebound will also add to this effect.

Tight At Mid Corner

Use stiffer right rear compression. This delays weight transfer onto the right rear.

If the chassis tightens up as the driver adds throttle, use stiffer right front rebound. This delays weight transfer to the rear.

Loose At Corner Exit

With heavier, faster cars or dirt track cars that use a locked left rear, use a softer rebound on the right front shock. This allows weight to transfer quicker from the front to the left rear. When combined with a softer compression shock at the left rear, weight transfers quickly from the front and settles at the left rear to tighten up the chassis. Both front shocks should be the easy-up type.

If the left rear is unlocked, use a softer rebound on the left front shock. This allows weight to transfer quicker from the front to the right rear. When combined with a softer compression shock at the right rear, weight transfers quickly from the front and settles at the right rear to tighten up the chassis.

Tight At Corner Exit

Increase the rebound damping at the left front. This keeps weight from transferring too quickly from the front to the right rear under acceleration.

Increase the compression damping at the right rear. This

If a car is tight at mid corner, use a shock which has a stiffer right rear compression valving.

The shock absorber controls how fast weight transfers onto and off of a corner of the car. If you use a stiffer shock in both compression and rebound at one corner, the weight will transfer slower to and from that corner. If the shock is softened in both compression and rebound, the weight will transfer quicker from and to that corner.

will slow weight transfer to the right rear.

Other Shock Tuning Tips

1) The shock absorber controls how fast weight transfers onto and off of a corner of the car. If you use a stiffer shock in both compression and rebound at one corner, the weight will transfer slower to and from that corner. If the shock is softened in both compression and rebound, the weight will transfer quicker from and to that corner.

2) In general, more rebound control and less compression control on the left side will make a car looser at turn entry. More left side tie down will keep the left side tires loaded further into a corner and lessens left side loading, freeing up the car at corner entry.

3) If the left rear is locked, softer compression control at the left rear will tighten up the rear of the car at corner exit by allowing weight to transfer to that corner quicker.

4) Softer rebound control in the front shocks will create better forward traction under acceleration off the corner by transferring weight to the rear quicker.

5) If the car skates across the race track when bumps and ruts are encountered, the shock absorbers are probably too stiff in their rate. Try using shocks one step softer.

6) If the shocks are too soft when bumps and ruts are encountered, wheel hop will occur. A wheel hop cycle occurs when the tire leaves the track surface, and then comes crashing back down.

Dirt Track Conditions and Shock Requirements

The most important thing for your understanding of choosing the correct shock is how to analyze the race track, and how to relate that to the different functions of shock absorbers. For example, you need to know which shock to put at the right rear when the track is really heavy and tacky, versus which shock to use there when the track is real dry/slick.

Heavy/Tacky Track

On a heavy or tacky track, the bite or traction is built into the track. There are not a lot of extra things needed to get the tires to grip. In fact, traction has to be taken out of the chassis to get the car to turn.

Shocks help the transitional weight transfer loosen up the chassis. Delaying or slowing weight transfer onto the right rear will prevent it from getting good side bite at corner entry. For a very tacky track, both rear shocks are one step stiffer to delay transitional weight transfer from the left rear to the right rear.

Using a tie-down shock at the left front can also be beneficial on a very sticky or tight track. This lets the left front drop down easily at corner entry and take weight off of the right rear. This makes the car want to pivot around the left front, which makes it easier to steer into the corner. The stiffer rebound valving delays weight transfer from the left front to the right rear.

Dry Slick Track

The track is really dried out. It is very hard and very slick, and it is difficult for the tires to get traction. You want weight to transfer very quickly from front to rear as the car accelerates off the corners to provide more rear traction. Both front shocks should have a softer rebound than compression to allow the car to lift the front end and transfer weight to the rear under acceleration.

Using a 2 compression / 5 rebound tie down shock, the 2 compression at the left rear lets the weight transfer to that corner quickly. The 5 rebound keeps that corner of the car tied down and stable at corner exit so it won't rebound.

Many racers will use a straight 2 valving shock on the left rear on a dry slick track. This is especially helpful at corner entry and mid-corner. The light 2 valving in rebound allows weight to transfer quickly to the right rear to provide better grip for that tire.

Tires & Wheels

Chapter 4

Tires are the most critical element influencing race car handling and performance. The tire contact patch size, how the tire grips the race track, and the sidewall flex are all factors that affect performance. The tire contact patches are the only interaction between the car and the track. Choices in tires, compounds, pressures, and chassis setup dictate how the tires are loaded dynamically, and make tires a very critical element.

Tire Dimensional Sizing

A typical quarter midget tire has a size of 11x7.10 – 6. The first number, 11, indicates the tire diameter in inches, 7.10 is the tire tread width, and 6 is the rim size (diameter). A 10x4.50 – 5 tire size would indicate a 10-inch tire diameter, a 4.50-inch tread width, and a 5-inch rim size.

Tire Compounds

The tire compound designates how soft or hard a tire's rubber is. Tire hardness determines how a tire's tread sur-

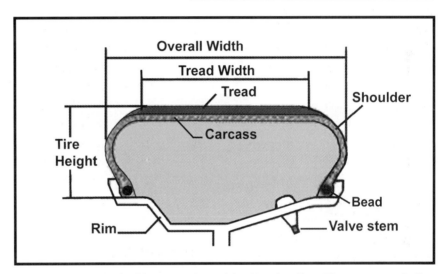

Terms connected with a quarter midget racing tire. The carcass is the main structural body of the tire. It is made of two layers of rubber coated nylon cord. The bead is made of a bundle of steel wires and forms the base of the carcass. The function of the bead is to form an air seal between the wheel and the tire, and keep the tire seated on the wheel. The tread is the area of the tire which makes contact with the race track.

The sizing of this tire indicates an 11-inch diameter, a 5.00-inch tread width, and a 6-inch rim diameter.

Racing Tire Durometer Hardness

Bridgestone/Firestone Durometer Hardness

YEX	YFA YHA	YGF YHB	YFF	YFH	YGC YHC	YBH YFG	YGJ
47	49	50	53	56	58	59	62

Soft (High Grip) — Hard (Low Grip)

Burris Racing Tires Durometer Hardness

SS-11 TX-11	SS-22 TX-22	SS-33	SS-44	SS-55
40	45	56	60	70

Soft (High Grip) — Hard (Low Grip)

Dunlop Racing Tires Durometer Hardness

DAM DBS	R6	DAH DBM	RH2	SL4
54	56	58	59	60

Soft (High Grip) — Hard (Low Grip)

Maxxis Racing Tires Durometer Hardness

HG5	WT4	A01	HG4	HT3 HG3
38	44	48	55	60

Soft (High Grip) — Hard (Low Grip)

face grips the track. A softer rubber compound grips a track surface better, and gets sticky at a lower operating temperature. Because of these factors, a softer tire also wears more quickly. A softer tire compound is generally chosen to provide better grip when the ambient air temperature and track temperature is cooler. A harder tire compound is generally chosen when the ambient air temperature and track surface temperature is higher. It takes a higher operating temperature to make a harder compound rubber get sticky and provide sufficient grip. A harder compound tire is more durable and wears better than a softer compound. Lower horsepower cars usually use harder compound tires. This is because the softer, stickier tires will provide too much grip and bog down the race car during cornering and acceleration.

A softer compound tire gives good adhesion (stickiness) but very poor durability (high wear). Softer, stickier tires also generate more heat. On abrasive track surfaces a softer compound can experience rapid and excessive wear. Softer compound tires work better on a cooler or slicker track surface. Harder compound tires work better on a hotter or more abrasive track surface.

A tire's hardness is affected by temperature, age, and the number of heat cycles it has endured. The hardness of a tire is one thing when it is new and unused, but it changes when these factors are applied. An instrument called a durometer is used to measure the relative hardness of a tire. A durometer measures the rubber hardness on a scale of 1 to 100, with the higher number representing a harder rubber compound. Lower numbers represent a softer, stickier tire.

Because temperature influences a tire's hardness, any time you measure a tire's durometer hardness, also check the temperature of the tire. When you want to compare the relative hardness of one tire to another, always check the temperature of each tire. If the temperature is different, the durometer readings will not be a valid comparison.

A tire's compound designation can be found on the sidewall of all tires. This tire is a Dunlop DBS, which is a softer compound tire.

Tire Heat Cycles

One heat cycle of a tire is the process of bringing the tires up to race operating temperature and then allowing them to cool. See "Racing Tire Break-in Procedure" later in this chapter for more information.

As tires experience more heat cycles, they lose their stickiness and get harder — both at ambient temperature and racing temperature. Tires that have more heat cycles on them will provide less grip. That is why new tires always provide better performance. Harder compound tires are less affected by heat cycles, and wear and grip more evenly for a longer time. Usually a racer can get many

One heat cycle of a tire is the process of bringing the tires up to race operating temperature and then allowing them to cool. As tires experience more heat cycles, they lose their stickiness and get harder, both at ambient temperature and racing temperature.

race days worth of use out of harder tires. Softer tires are more responsive to heat cycles and lose their grip sooner as they are heat cycled.

Tire Softeners & Chemical Treatment

Chemical tire treatments alter the characteristics of a tire. They are used to enhance the grip of a tire, both during qualifying and during a race. Some racing organizations allow this, and some do not. Be sure to check your rules before using any type of tire softening treatment.

Some of the chemical treatment products enhance tire grip throughout the course of tire life, thus increasing the useful life of a tire. By using a chemical treatment to restore an older, hardened tire to a new pliable condition after many laps of racing, tire life is stretched.

There are a variety of different tire softeners or solvents on the market that racers use to soften their tires for qualifying. A tire softener solution is painted onto the tread surface of each tire before qualifying. This solution makes the tires softer, which simulates having run the tires a few laps and bringing them up to operating temperature. This gives the tires more grip for the first couple of laps. After a few laps, the prepped tire surface will wear off and the tires will perform like normal.

Be careful when applying any type of tire chemical treatment, and read the directions before using a product. Chemicals used in some of the products are dangerous, and should not touch your skin. Wear gloves and eye protection as a precaution, and do not breathe the fumes of any of the solutions. The solution should be applied to tires before going to the race track so it has a chance to thoroughly dry.

Track Tire Rules

Some racing associations specify the use of a track tire or "spec" tire. This is one specific tire that all competitors must use, instead of having an "open" tire rule where any tire can be used. The track tire compound is usually hard, which is intended to lengthen tire life and thus cut operating costs. This harder tire compound also makes chassis setup more difficult and challenging because a harder rubber compound provides less tire grip on the track surface.

Tires For Qualifying

If you have a choice of tires that can be used for qualifying, you want to use a softer compound for qualifying than you would for a race. This is true for both paved and dirt tracks. If using a softer compound is against the rules, or out of the

On both paved and dirt tracks, the tires must be cleaned off before qualifying or racing, Scrape the tires to remove all built-up rubber and debris.

Using a file to groom the tires is a great idea. It removes all rubber debris and dirt and gravel, and prepares a clean, even surface for the tire to grip the track.

question monetarily, use the freshest tires you have for qualifying. This helps to produce better grip in a short run. If you use new tires for qualifying, be sure to scuff them for 3 to 5 laps first, well before qualifying, then let them cool. Brand new tires, without being scuffed, will be slick and not provide proper grip.

On both paved and dirt tracks, the tires must be cleaned off before qualifying. If you are racing on a paved track, scrape the tires to remove all built-up rubber and debris. Get the tires clean so they can grip the track. If you are racing on dirt, wash the tires. Again, this cleans the surface so the tires can get a better grip.

Tire Inflation Pressure

Air pressure does three very important things that change the character of a tire and its traction capability. It changes the size of the tire contact patch, or footprint, it changes the lateral stiffness of the sidewall, and it changes the spring rate of the tire. (The lateral stiffness is a factor of the total spring rate of the tire.)

Proper tire pressures are a compromise between being low enough to provide good traction and grip, and high enough to support the shape of the tire so it does not distort laterally or roll under.

The correct tire pressure also allows the tire to use the entire tread face for traction, and prevents overworking the tire which can cause excessive heat build-up.

A higher inflation pressure decreases tire grip on the track surface because the contact patch area is smaller, the tire spring rate is stiffer, and the lateral stiffness is harder.

A lower inflation pressure increases tire grip on the track surface because the contact patch area is larger, the tire spring rate is lower, and the

Because a tire is a flexible structure, it functions as a spring. Because air pressure is what supports that flexible structure, air pressure determines the spring rate of the tire. Adding more air pressure makes a tire's spring rate stiffer.

lateral stiffness is softer. When adjusting tire pressures, be careful to keep the pressures within a reasonable range from low to high to prevent erratic handling and tire damage.

Because a tire is a flexible structure, it actually functions as a spring. And because air pressure is what supports that flexible structure, air pressure also determines the spring rate of the tire. The tire's spring rate adds to the suspension spring's rate at each corner of the car to give an overall spring rate. So, adding or subtracting inflation pressure in the tire is going to add or subtract spring rate.

Adding more air pressure makes a tire's spring rate stiffer. Subtracting air pressure makes it softer. So, for example, when less air pressure is used in the right rear tire when the track gets slick, it is the same as using a slightly softer right rear spring. And what that does is tighten up the rear of the car, counteracting a loose condition.

Less air pressure creates more tire patch, or tire footprint, which in turn gives better tire grip. Lower air pressure is used in situations where more bite is required, or bite is hard to get, such as on real slick tracks. One to two pounds of air in the right rear tire has a significant effect on handling. Taking out two pounds of air reduces spring rate, allows the sidewall to be more flexible, and increases tire contact patch size.

Lowering the tire pressure will create a larger tire footprint on the track, but it will also create more tire heat. When the inflation pressure is lower, the tire will operate at a higher temperature. This occurs because the cords of the tire body have much more friction with less inflation.

Optimum tire pressures are difficult to generalize because there are so many factors that influence the pressures and proper handling. Those factors include the weather and temperature, track surface condition, track length and banking, and chassis balance.

Dirt Track Air Pressure

On a dirt track, when the track is heavy, you want to keep the tire from rolling over and flexing the sidewall more. You don't need a lot of tire patch here because the bite is built into the race track. More air pressure is used on heavy tracks to decrease tire patch, and to keep the tire sidewall more rigid. Wetter, tackier tracks require more tire inflation pressure. Drier, slicker tracks require lower tire inflation pressure.

While lower air pressure is beneficial on a dry slick track to gain more tire footprint, be careful not to run too little pressure. This will let the sidewall flex too much and the tire will roll under. Also, if too little inflation pressure is used, the center of the tire can be distorted upward into a concave shape because there is insufficient pressure to support the tire. This reduces center tread contact with the track and reduces footprint size and grip.

Less air pressure creates more tire patch, or tire footprint, which in turn gives better tire grip. Less inflation pressure makes the tire slightly more flexible, which gives it a little more traction. But if a tire is too low on inflation pressure on a paved track, the center section of the tire will not be loaded properly and the tire will begin losing grip. The ultimate inflation pressure is a very narrow range. That is why it is so important to monitor tire inflation.

Keep accurate records of how much air pressure builds up in each tire over a certain number of laps. Keep a record of track surface conditions, track temperature and ambient air temperature as well, because all of these factors influence how much air pressure builds up and how quickly. Generally, the majority of pressure build-up will occur in the first ten laps of racing. Lowering the starting pressures will allow you to control the pressure build-up in the tires.

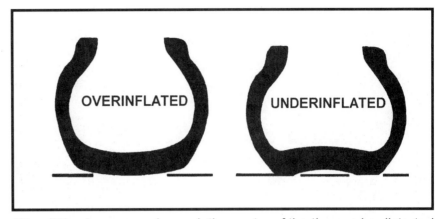

If too little air pressure is used, the center of the tire can be distorted upward into a concave shape because there is insufficient pressure to support the tire. Likewise, overinflating a tire will distort the center of it and the full tread face will not contact the track.

Having accurate records and information will help you decide on the starting air pressures before each race. The goal is to end a race with the optimum air pressure in each tire.

Factors Affecting Tire Pressure

Understeer/Pushing
When a car is pushing, increase the rear tire pressure by 2 to 3 PSI. Higher rear inflation pressure makes the tire and sidewall stiffer, which decreases grip. It also increases the spring rate of the tire, which loosens up the chassis.

Oversteer/Loose
When a car is loose, decrease the rear tire pressure by 2 to 3 PSI. Lowering the rear inflation pressure makes the tire and sidewall more flexible, which improves tire grip. Better rear tire grip counteracts oversteer. When adjusting tire pressures for oversteer or understeer, make the pressure adjustments in the rear tires only.

Using Nitrogen
One factor that affects heat build-up and retention, as well as pressure build-up, is the moisture content in the air used to inflate the tires. Moisture expands when subjected to heat. Using an ordinary air compressor, the amount of moisture pumped into a tire is the same as the humidity content in the air. On high humidity days, the moisture content is equally high. This puts more moisture inside the tire and creates pressure build-up and heat build-up.

One of the ways to combat this problem is to use nitrogen to inflate tires. Go to your local welding supply store and purchase a bottle of nitrogen.

Every time the race car comes into the pits, take the time to check the tires. Check tire pressure, tire stagger, tire wear, the pattern of tire wear, and the temperatures.

It will have a lot less moisture content than ambient air. Bottled nitrogen has a water content of ten parts per million whereas compressed air (with moderate humidity) is 500 parts per million. The important point is that if you always use the same nitrogen source, the moisture content will always be the same and thus pressure build-up and heat build-up will always be consistent and predictable. It won't be erratic like relying on the moisture content of ambient air pumped in by an air compressor.

Checking Tire Pressure

It is extremely important to record the cold tire pressures just before the car goes onto the track so these is a base of comparison for hot tire pressures. Record the starting tire stagger as well.

Every time the race car comes into the pits, take the time to check the tires. Check tire pressure, tire stagger, tire wear, the pattern of tire wear, and the temperatures.

Check the tire pressures first, starting at the right rear, then the left rear, and then right front and left front. Check rear tire stagger right after rear pressures are taken. If stagger has changed, it should relate to differences in tire pressures Check temperatures and pressures quickly before the tires cool appreciably. This information is vitally important.

Make this a regular procedure so you can stay in touch with what the chassis and tires are doing. Keep good notes.

Checking Tire Temperatures

A useful instrument to help you track tire pressures and chassis performance is a tire pyrometer. It is an electronic instrument which gives a temperature reading when its probe is inserted into a tire surface.

Tire temperatures are read with the pyrometer at three positions across the face of each tire and are recorded on a sheet of paper in the manner shown in the accompanying illustration (next page).

By comparing the temperatures across the face of each front tire, it can be determined if each tire has too much or not enough camber, or if inflation pressure is correct. If the temperature in the middle of the tire is higher than the inside and outside edges, that indicates the tire is over-inflated. If the middle temperature is lower than the inside and outside edges, that indicates inflation pressure is too low.

Comparing the average temperature (of all three positions) of the right front tire to the average of the right rear will determine if the chassis is tending toward understeer or oversteer.

Tire temperatures can change very quickly and cool off very fast. Take the temperatures very quickly, starting at the right front and working around the car in a circular motion to the right rear, ending up at the left front. See the *"Track Tuning & Adjustment"* chapter for more details on tire temperature analysis.

Tire Stagger

Stagger is the difference in inches of tire circumference between the left rear and right rear tires. Stagger is required because two tires on the opposite ends of a solid axle are running on two different radii through a turn. The outside tire must travel a greater distance on a wider arc than the inside one. This is accomplished by the outside tire being larger in circumference than the inside tire so it runs at a slightly faster speed. The tighter the radius of a corner, the larger the circumference of the outside tire must be in relationship to the inside. The tighter the race track, the more stagger required.

Track conditions also dictate the amount of stagger used. If you are having a difficult time making the race car turn, stagger can help accomplish it.

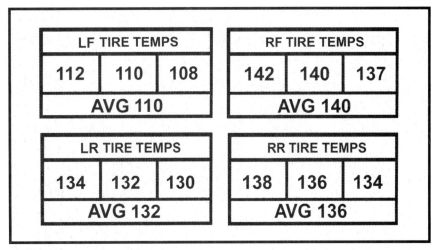

Record tire temperatures in a manner such as this so they are easy to visualize and compare. The average temperatures are important for determining handling balance.

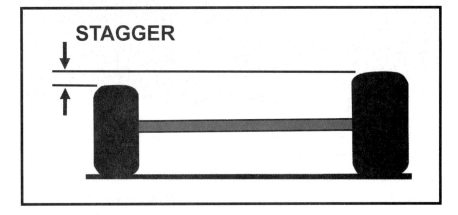

Stagger will make the right rear overdrive the left rear and drive the car in a tighter arc.

The width of the wheel on which a tire is mounted can also affect stagger. See "The Effect of Wheel Width" at the end of this chapter.

Growing A Tire

In some cases it is desirable to increase the circumference size of a tire to obtain a desired amount of stagger. This can be accomplished by increasing the inflation pressure of a mounted tire, and letting the tire set in the sun for a couple of hours. Do not use any more than 35 PSI inflation pressure. Measure the circumference periodically to make sure the tire does not grow beyond your goal. When the tire has reached the desired size, deflate the tire and cool the tire with cold water.

Reading Tire Surfaces

Reading racing tire surfaces is something of an art form. By reading the wear pattern on the surface of the tire, a racer

A tire with optimal tire wear will show an even grainy pattern across the rubber surface. This tire indicates excellent grip.

is better able to determine handling characteristics of the vehicle. Combined with tire temperatures, sound judgments can be made about changes that need to be made to improve handling characteristics. Often the ability to read the wear patterns is more valuable than reading tire temperatures. This is partly due to the characteristics of the race track where you are competing. Tire temperatures are more indicative of the last turn or two on a given race track. However, tire wear patterns will be indicative of what goes on all the way around a race track. -Each and every time your car comes in off the race track, you need to look at the tires and analyze what they are telling you. Tires are an important part of the complete handling package. You are not going to win races if you don't understand tires.

Normally a new tire's tread surface is a dull blackish gray color. A tire with optimal tire wear will show an even grainy pattern across the rubber surface.

If a portion of the tread surface appears shiny or glazed, this is normally an indication that the tire compound is too hard and it has sealed over. If a portion of the tread surface shows no wear signs at all, this also indicates the tire compound is too hard. A softer compound tire is required.

If rubber is tearing off the tire or shredding off the tire blocks (on a dirt treaded tire) like a soft eraser, this indicates the compound is too soft.

If a tire is blistered on the inside shoulder and cold on the outside, this indicates there is too much stagger being used, or the tire is overinflated. If a tire is hotter and/or blistered on the outside edge, this indicates the air pressure is too low.

Because tire stagger creates negative camber in the right rear tire and positive camber in the left rear tire, the inside edge of the rear tires will generally wear more than the rest of the tire. (The inside edge of each tire is defined as the edge facing the infield.)

It is important to look at what is happening in the middle of the tire. The wear pattern should show a fine grain across the face of the tire. The very inside edge of the right rear tire may exhibit accelerated wear along the shoulder, but this is acceptable as long as the rest of the tire looks normal. The inside edge of the right rear along the shoulder will run hotter than the face of the tire because of stagger.

A tire that is working well will have a fine grain appearance, and it will be smoother and more uniform than a tire that is running too hot.

If a tire is too cool and the

tire face shows very little or no graining pattern, the tire compound is too hard for the application.

If the graining pattern shows excessive tearing and pulling on the face of the tire, the compound being used is too soft.

If the center of any tire shows excessive wear, it indicates too much inflation pressure.

If a tire surface appears quite grainy and some rubber chunks are being pulled off the tire, then the tire is building up too much heat and has too much grip. The higher the tire compound temperature, the lower the tear strength of the tire surface. In other words, you can get the tire surface overheated to the point that track surface abrasion will pull chunks of the tire surface off.

Developing the ability to read tires is important. One method which is effective to develop this quality is to study the tires of your top competitors. Compare the wear patterns and grain patterns on competitors' tires relative to their performance on the track. By practicing this, you will become adept at reading the tires on your own race car and subsequently be able to improve handling and increase the performance of your car.

Racing tires go through a "final cure" in their first heat cycle. Race tire break-in should involve gradually working the tire up to racing speed for a few laps until the tires reach normal operating temperature, then bring the car in and allow the tires to cool.

Racing Tire Break-In Procedure

Racing tires are very influenced by heat the first time they are run. They go through a "final cure" in their first heat cycle. At this time the resins in the compound stabilize, the rubber cures, and the tire assumes its final dimension. During the first heat cycle of the race tire, it runs hotter than during its normal operating range.

Race tire break-in should involve gradually working the tire up to racing speed for a few laps until the tires reach normal operating temperature. They should not be run for more than a couple of laps at normal operating temperature, nor should they be abused or run hard the first time. If they are abused, they will change compound hardness or blister, and be ruined as a suitable racing tire. Have the car race-ready, and set the tire inflation to normal starting pressures. Build up speed gradually on the track, then run two to three laps at racing speed. Then bring the car into the pits and allow the tires to cool. That establishes the first heat cycle for the tires.

The biggest mistake racers make is trying to get racing tires too hot too quickly. Tires should be broken in slowly,

Keep the tires wrapped in their factory wrapping until they are used. This will slow down the deterioration process.

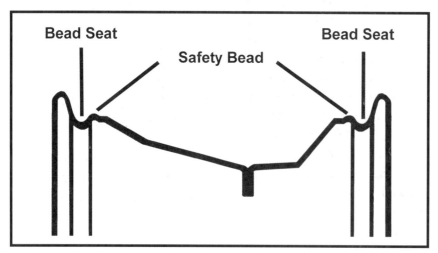

When mounting a tire, the tire beads must be seated in the bead seats of the rim before inflating the tire.

rocks and gravel, and other small pieces of debris from the track. Make sure they are thoroughly clean before you go racing.

Tire Use And Aging

The newer a tire is after it is manufactured, the better it will grip the track. The longer a tire sets around before it is raced, the harder the tire rubber gets and the less grip it will produce.

When a tire is used over and over again, it gets harder and loses its grip of the track surface. The tire has been heated and cooled so many times that the compound oils are sweated out and the tire gets hard. After repeated use and heat cycles, it is very inefficient to grip the track properly.

To slow down the deterioration process, store tires in a cool dark place. Store tires out of the sun. Ultraviolet light will deteriorate the tires quickly. Keep the tires wrapped in their factory wrapping until they are used.

Tire Mounting

Mounting and inflating a tire can be very dangerous. Be sure to follow recommended safety procedures. This is a job for experts with proper knowledge.

Use a proper lubricant on the tire beads and rim to mount the tire. Instead of using soapy water, use a product

with proper inflation, running only two to three laps at racing speed, and then allowing them to cool.

After the tires have cooled, clean them thoroughly. Hot tires will pick up tire marbles (small bits of rubber), small designed for tire mounting, such as Tire Snot (available at most kart shops). This product dries soon after the tire is installed, whereas water left inside the tire will cause pressure build-up when the tire gets hot. Another product to use as a tire bead lubricant is Windex.

Make sure the tire beads have seated on the rim before inflating the tire. Always inflate the tire in a safety cage or other restraining device. Use an air hose with a clip-on air chuck. Do not stand over the assembly while the tire is inflating.

Never inflate a tire beyond 40 PSI to get the beads to seat. If they don't seat, deflate the tire, dismount it, lubricate the beads and rim again, and start over.

Once a tire is mounted, check for air leaks. Apply soapy water around the bead seat areas and valve stem, and look for bubbling.

Also be sure to check the tire for any type of damage which might cause on-track tire failure – tire cuts, cracks or foreign objects being embedded into the tire. In the interest of safety, any damaged tire should be discarded.

Tire Selection For Dirt Tracks

Racing success on dirt is dictated by tire selection. Tire compounds and tread patterns influence tire traction.

The proper compound has to be selected to provide the right grip and wear characteristics. No matter what adjustments are made to the chassis, if the compound is not correct, the optimum grip will not be achieved.

If a tire shows a lot of wear and feathering, it is too soft. A harder compound should be used. When a track is hard, slick and has very little grip, a harder tire is required. How hard the tire should be depends on how abrasive the track surface is. A softer tire will give more grip and side bite. But if the track is abrasive, a softer tire will wear very fast. In many cases, dirt oval racers use right side tires that are a harder compound than the left side tires.

An alternative when a track gets hard and slick is to use a slick tire rather than a grooved dirt tire. This works better because a slick puts more rubber on the track for better grip.

Sipe cuts on a tire make the tire gain heat faster, which builds tire grip quicker.

"Siping" a tire is the process of making small razor cuts across the tire blocks. A siping tool uses a sharp blade which makes a small cut to create more track gripping edges across each row of blocks, and make the blocks more pliable. Sipe cuts allow heat to build in the tire quickly and lets a harder tire start working earlier. Tire siping is most helpful

"Sipe cuts" on a tire make the tire gain heat faster, which builds tire grip faster. Siping can be used on dirt or paved track tires.

on a dry slick track where the tire traction is very delicate when the tires are cold. The sipe cut should not be much more than just a scratch on the block surface. All it has to do is add extra grip for a few laps until the tire comes up to operating temperature. Too deep of a sipe cut will cause the rubber to pull loose and roll under the tire, making traction very loose.

If the track is starting out a little soft and will get harder, and the tire will be initially too hard, the sipe should be made to a depth in the block that will get worn away by the time the track gets harder. How far is that? Experience is the best teacher, but be sure to pay close attention to the amount of tire wear at all times.

It is an art form to learn how

to do proper sipe cuts. As a general guideline, if the sipe cut still shows up on the face of the tire at the end of a feature race, then you cut it too deep. Practice and observation will help you learn.

Selecting The Right Dirt Tire Compound

You can choose from a variety of compounds from very soft to very hard. The proper tire hardness is determined by the track conditions. On dirt tracks, conditions can change greatly during the night from the beginning of hot laps to the feature race. Therefore you must choose a compound that will be effective at the end of the race. It may not be as fast early in the evening, but if you choose one that is too soft, you won't be in the ball park at the end.

If you question whether your compound choice will be too soft or too hard, always lean toward being a little too hard rather than too soft. A harder compound tire will be there at the end, whereas a softer one may give up.

In most cases, softer tires are used for qualifying, heat races, wet tracks, or tracks that don't have a lot of abrasion. The high moisture content of the track surface keeps the tires cool. Softer tires can also be used on harder tracks that don't build a lot of heat in the tires, or that have loose surface dirt. You will also notice that softer tires wear at a faster rate and will tear or rip easily. This is something you have to take into consideration when you use a softer tire. If the track surface is abrasive or has rocks in it, a harder compound is a must. You have to learn how long the tire is going to do its job on a particular race track condition.

Track temperature will also affect tire compound choice. A cooler track will allow a tire to operate cooler, so a softer compound can be used.

Harder tire compounds are required when the track gets harder and the tires work on the track surface. This is because tires on a hard track build up more heat, and more friction is generated between the tire and the track surface.

Higher speed tracks will develop more tire heat and thus require a harder compound tire. Higher banked tracks generate more tire heat as well, and require a harder compound.

Tracks that have a lot of rocks or pebbles promote wear at a more rapid pace, so take this into consideration when choosing a tire compound. This type of track is going to be abrasive, so this will require a harder compound tire. A track that packs down and develops an asphalt type of surface will also require a harder compound. When a track gets hard and black, a tire with the least amount of grooves in it is the most effective. This is because having the greatest amount of rubber tire surface produces the greatest amount of grip on the track.

A race track that is hard but yet doesn't get abrasive or build a lot of heat in the tires might require some tire siping. A track like this will sometimes require a hard compound tire. This in turn might cause a tire to slick over or glaze. Siping adds flexibility to the tread blocks, which helps prevent the glazing problem.

Watch track conditions closely. If you are not familiar with how a track might change, find out from people who race there regularly. When you choose your compound, whether it be hard, soft, or somewhere in between, if there is a doubt about which one to use, go with the harder of the possibilities.

Many times the correct compound for a race is just an educated guess, based on all the factors and the expected condition of the track. Only after the race is over do you know if the guess was right or not. That's a big part of the challenge of racing on a dirt track! Be sure to keep good notes so you have experience to draw on in the future.

Wheels

Quarter midget wheel sizes are quoted in sizes such as

5"x5" or 8-1/2"x6". The first number is the wheel width, and the second number is the wheel diameter. Normally 5-inch diameter wheels are used on the left side and 6-inch wheels are used on the right.

Typically, quarter midgets will use 5"x5" wheels on the left front, 5"x5", 5-1/2"x5", or 6"x5" wheels on the left rear, 6"x6" or 6-1/2"x6" wheels on the right front, and 8"x6" or 8-1/2"x6" wheels on the right rear.

A wheel should be ½-inch to 1-inch wider than the tread width of the tire mounted on it. For example, a 6-inch wide tire should be mounted on a 6 ½-inch or 7-inch wide wheel.

Most quarter midget wheels use the American bolt pattern (or standard bolt pattern) which has a 1.75-inch register hole and three bolt holes on a 2.5-inch bolt circle.

Wheel Backspacing

Wheel backspacing is the distance from the hub mounting center of the wheel to the inner edge of bead seat (NOT all the way out to the outside of the rim). If that measurement is 3 inches, then the wheel has a 3-inch backspacing. One-piece wheels are available in 2 through 4-inch backspacing, in half-inch increments. Typically, a left front wheel uses 3-inch backspacing, a left rear uses 2-inch backspacing, a right front uses 4-inch backspacing, and

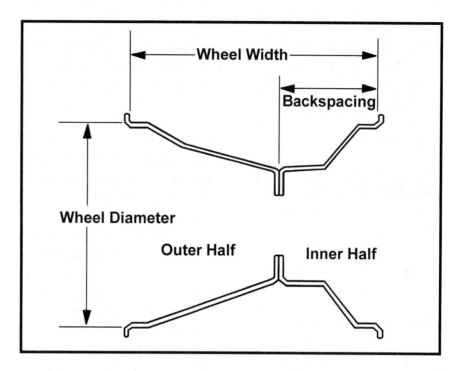

a right rear uses 3 ½ or 4-inch backspacing.

With two-piece wheels, the inner half size is the wheel's backspacing. For example, a "3 on 2" or "3 x 2" wheel uses a 3-inch inner half and a 2-inch outer half, resulting in a 5-inch wide wheel with a 3-inch backspacing.

The Effects of Wheel Width

The circumference of a tire varies in response to the width of the wheel it is mounted on. The wider the wheel it is mounted on, the smaller the circumference of the tire. For example, a 6-inch tread width tire might measure 34.75 inches in circumference mounted on a 7.25"x6" wheel. But when the same tire is mounted on a wider 7.75"x6" wheel, it may measure 34 inches in circumference, or when mounted on an 8.25"x6" wheel it may measure 33.5 inches in circumference. These numbers may vary slightly depending on individual tires.

Using this information, it can be seen that mounting a particular tire on a wider or narrower wheel can be used as an effective means of changing tire stagger. A ½-inch wider wheel may decrease tire circumference by ½-inch to ¾-inch.

The width of a wheel can also influence the handling characteristics of a race car. The width of the wheel affects how much the tire sidewall will flex. A narrower wheel width allows more sidewall flex, which increases tire traction. Mounting a tire on a wider wheel decreases side-

Two-piece wheels generally use a 6-bolt pattern for better clamping of the center section, whereas one-piece wheels use a 3-bolt pattern.

wall flex and makes the sidewall stiffer. This stiffer sidewall provides less tire grip. So, for instance, mounting rear tires on slightly wider wheels would tend to make the handling of the car looser. On the other hand, mounting the rear tires on slightly narrower wheels would provide more sidewall flex and increase tire grip, which tightens the chassis.

One-Piece Vs. Two-Piece Wheels

Quarter midgets use both one-piece and two-piece wheels. However, many chassis manufacturers are standardizing on one-piece wheels. One-piece wheels are lighter weight than two-piece wheels, which means quicker acceleration. In addition, one-piece wheels are stronger than two bolted halves, tires are easier to mount on a one-piece wheel, and the one-piece wheel does not need a seal.

Two-piece wheels can be more cost-effective than a one-piece. If one half of a wheel gets bent or damaged, just the half can be replaced, and not the entire wheel.

Wheel Inspection

The race car should undergo regular maintenance and inspection after each day of racing. Part of this maintenance should be wheel inspection. Check the wheels for damage and wear after each race, and every time a new tire is mounted. Also check wheels immediately if the car has made contact on-track with other cars or the wall. Damaged wheels should be replaced. Bent wheels will accelerate tire wear and fatigue suspension parts. Bent wheels cannot be repaired because the damage has compromised the structural integrity of the wheel.

Check the lug nut holes for wear or elongation, This is caused by loose lug nuts. If you find this condition, throw the wheel away because this will prevent the lug nuts from properly clamping, and the lug nuts will come loose easily. Also check for fatigue cracks around the lug nut holes. When these develop, it is time to replace the wheel.

Chapter 5

Chassis Setup & Alignment

The basics of chassis setup are very simple: get it bind free and square. This means taking the time in your shop to measure and adjust the car so you are starting with a proper foundation which can be relied upon to make adjustments to the chassis and have predictable results. The biggest part of being a successful racer is all the preparation that goes into the car before going to the race track. Take the time to make sure the chassis and suspension components are square – straight in all planes when measured against a common reference point, and make sure the ride heights and corner weights are correct.

The biggest part of being a successful racer is all the preparation that goes into the car before going to the race track. Take the time to make sure the chassis and suspension components are square – straight in all planes when measured against a common reference point, and make sure the ride heights and corner weights are correct.

The Baseline Chassis Setup

By doing the baseline chassis setup, you will get the axles squared to the chassis, the suspension components properly aligned, and the basic weight distribution set along with chassis ride heights and rake and tilt. This becomes the foundation that all other chassis adjustments are built on.

As you take measurements of the car, make it comfortable for yourself. Place the car up on a work stand or a pit cart.

Start with the chassis and suspension fully assembled, but without wheels and engine installed. This makes the adjustment procedure easier without components being in the way. Set the car in front

and rear alignment blocks.

Square The Rear Axle

The first step is to square the rear axle. The rear axle must be set 90 degrees to the chassis centerline. To square the rear axle, use a frame cross member which sets behind the driver's seat as the squaring reference point. Measure from the back of the cross member

Chassis Setup & Alignment

To square the rear axle, measure from the front of the axle to a chassis crossmember that is perpendicular to the car's centerline. When measuring to the axle, place a steel square against the front of the axle and measure to the square. The measurements on both sides of the axle should be the same.

to a steel square setting against the back side of the rear axle. Using a steel square helps to make the measurements more accurate. Set the square against the axle and measure to the square. Take this measurement on each side of the chassis. Be very accurate with the tape measure. These two measurements should be equal.

The above method assumes that the chassis crossmember being used as a reference point is absolutely perpendicular (90 degrees) to the car's centerline. To double check the accuracy of this, use the steel carpenter's square placed against the chassis crossmember and the string marking the centerline of the chassis. The square will quickly show you if the crossmember is perpendicular.

Adjusting The Rear Axle To Make It Square

If the rear axle is not square, it is adjusted by shortening or lengthening the upper and lower radius rods on one side of the car. If the axle is set back on the right side, for instance, shorten the two right side radius rods. The upper and lower radius rods must always be shortened or lengthened an equal amount to avoid angling the birdcage. Adjust the length of the radius rods until the axle is square.

Leveling The Birdcages

A birdcage is the axle carrier housing that is attached to the chassis with radius rods. There is one on each side. The birdcages must be leveled, or timed, to prevent roll steer during cornering.

Leveling or timing a birdcage

Chassis setup and alignment is always done with the car setting in front and rear alignment blocks.

Chassis Setup & Alignment

If the rear axle is not square, it is adjusted by shortening or lengthening the upper and lower radius rods on one side of the car.

A birdcage is the axle carrier housing that is attached to the chassis with radius rods. There is one on each side. The birdcages must be leveled, or timed, to prevent roll steer during cornering.

means getting it absolutely vertical so that the upper and lower link attachment bolts are perfectly aligned. This is accomplished by adjusting the length of the upper and lower radius rods.

Set a steel square or other vertical alignment straight edge next to the birdcage, and measure horizontally from the center of the top and bottom rod end attachments to the square. Both measurements should be equal. If they are not, loosen the jam nuts on the links and turn the links until the birdcage is absolutely vertical. Repeat this procedure on the other side.

Be sure that the axle is not pulled forward or rearward when making adjustments for birdcage timing. Adjust both the upper and lower links (one in, one out) until the birdcage is squared. Once the jam nuts

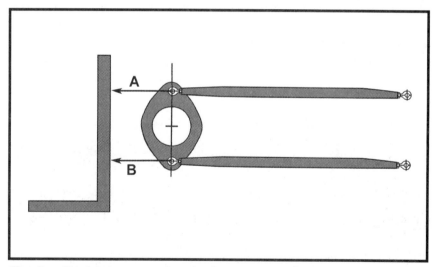

To align the birdcage, set a steel square or other vertical alignment straight edge next to the birdcage, and measure horizontally from the center of the top and bottom rod end attachments to the square. Both measurements should be equal.

This type of birdcage, with a flat vertical back, is much easier to square. Just place an angle finder or Smart Level against the rear of it, and adjust the links so the back is perfectly vertical.

To square the front axle, measure from the back of the front axle to the front side of the rear axle on each side. Use a steel square set against each axle to help make accurate measurements.

on the radius rods are tightened after adjusting the timing, double check the axle squaring measurements one more time to be sure nothing moved.

Squaring The Front Axle

Once the rear axle is square, you can use that as a known reference point to square the front axle. The front axle is squared now in the alignment procedure so that other components can be squared and aligned. After everything in the front is aligned, axle lead can be added (more on that later).

To square the front axle, measure from the back of the right side of the front axle to the front side of the rear axle. Use a steel square set against each axle to help make accurate measurements against round surfaces. Make sure there is no slack in the tape. Even 1/16-inch can make a difference in properly aligning the chassis. Then measure in the same manner on the left side. When the left and right measurements are equal, the front axle is squared to the rear axle. If the left and right side measurements are not equal, shorten and/or lengthen the front axle attaching radius rod on the left side to achieve an equal distance.

Setting the Panhard Bars

The Panhard bar height

establishes the roll center height both at the front and the rear suspension. Most race car manufacturers have designed their chassis to operate properly with the Panhard bars set in a specific location. For example, they may specify it be set at a 4-degree angle, or set level with the axle, or set in the third hole up in the frame bracket, etc. Start with the baseline recommendation from your car maker. You have to know what the starting point is for your car, or it may affect the handling performance.

An initial starting point for the front Panhard bar mounting height is the frame end of the bar ½-inch higher than the axle centerline. This makes the Panhard bar parallel with the front axle as the chassis rolls during cornering. This is a good starting point unless your chassis maker has another starting specification. The front Panhard bar can be moved up or down within a certain range to adjust for handling conditions.

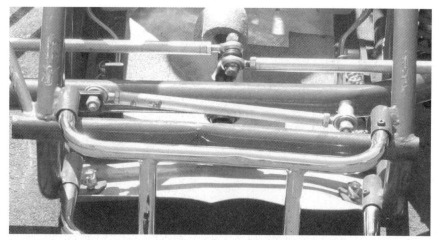
An initial starting point for the front Panhard bar mounting height is the frame end of the bar ½-inch higher than the axle centerline, or whatever the chassis maker specifies.

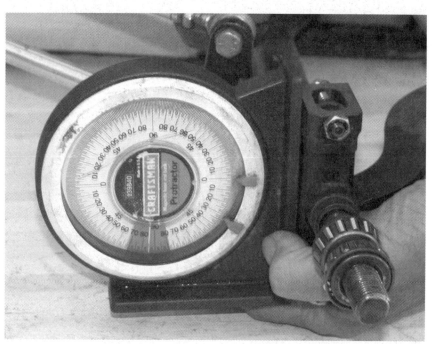
Set the caster by placing an inclinometer (angle finder gauge) against the back of the bracket which holds the spindle on the right front. Adjust the right side radius rod lengths to produce the desired angle.

Setting Caster

Once the front axle is squared, set the desired positive caster using the right side radius rods. Positive caster sets the top of the kingpin tilted rearward. For the initial setup, use 5 degrees of positive caster at the right front. In most cases, the front axle is built with 3 degrees to 5 degrees caster difference (depending on axle manufacturer). This means that when the right front is set to 5 degrees positive caster, the left front is set at 0 to 2 degrees positive caster. The positive caster split helps the car turn into a corner easier. Five degrees of right front positive caster is a good trade-off between straight line stability and steering effort during cornering. More positive caster helps promote

Camber can be measured using an inclinometer placed against the rim face. Make sure the rim is perfectly flat and it does not have any dents or dings.

Negative camber is set by moving the top of the spindle inward toward the car.

makes it difficult for a driver to drive smoothly.

The normal range of right front caster is between 4 and 6 degrees positive. Unless you know for certain how much caster you want, set it at 5 degrees positive, then fine tune the adjustment with track tuning and driver feedback.

Set the caster by placing an inclinometer (angle finder gauge) against the back of the bracket which holds the spindle on the right front. Adjust the right side radius rod lengths to produce the desired angle. Mark the rod end shank threads with a broad felt tip pen. This helps when setting the caster to know how much each rod has been moved. To avoid getting the axle out-of-square while setting the caster, move the upper and lower radius rods in equal amounts. For example, move the upper rod in one turn and move the lower rod out one turn.

After setting the caster, double check the front axle squaring. If the caster setting has moved one side of the axle, readjust the left front radius rod once again to square the front axle to the chassis.

Setting Camber

Camber is the inward or outward tilt of a tire at the top relative to vertical at the center of the tire. Zero camber is true vertical, negative camber is the tilt at the top of the tire toward the center of the car,

straightline stability (less wander). But the more positive caster a car has, the more physical effort required by the driver to steer the car during cornering. Less positive caster reduces steering effort, but the reduced effect many times and positive camber is the tilt of the top of the tire away from the center of the car.

The proper amount of camber maximizes the tire contact patch on the track. Optimum camber puts the entire tread on the track for even wear and grip. The normal range of camber is positive 1-degree at the left front, and negative 1 to negative 3 degrees at the right front.

To set the camber, place a race-ready car on a flat level surface. Make sure the front wheels are straight ahead. Attach a camber gauge to the spindle and level the bubble on the gauge. This will give you the current camber setting. Then set the camber as you watch the gauge. Exactly how the camber is adjusted is dependent on the chassis manufacturer and the system they use. There are several different methods utilized by various manufacturers. Refer to your car maker's instructions.

Initially, set the right front camber to negative 2 degrees (top of tire tilted inward toward the center of the car), and set the left front to positive 1 degree (top of tire tilted outward away from center of car). Track testing and tire temperatures will help to determine the optimum camber setting.

The amount of negative camber used at the right front depends on the class. Faster classes with larger, heavier drivers will require more nega-

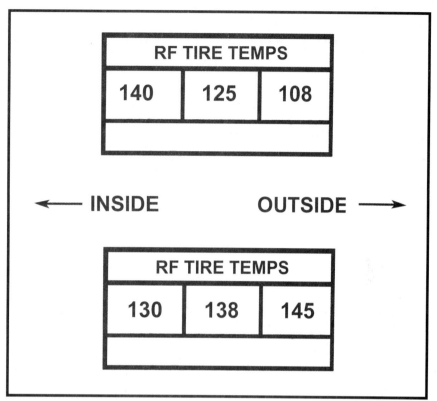

The top set of right front tire temperatures indicates too much negative camber. The bottom set of temperatures shows too much positive camber.

tive camber because these elements make the right front tire roll under during cornering. These cars enter the corners faster and experience more overturning moment, which causes more tire tread face deformation. On the other hand, slower classes with lighter weight drivers produce very little tire tread deformation, so camber can be initially set at 0 to degrees.

After the car has run a few laps on the track, use a tire pyrometer or heat gun as your guide for camber setting. If the inside edge of the right front tire is 10 degrees or more warmer than the outside edge, the car has too much negative camber (the tire is running more on the inside edge). Take some camber out. If the outside edge of the right front tire is 10 degrees or more warmer than the inside edge, the tire is rolling over and you need to add some negative camber. Make camber changes in 1-degree increments.

See the *"Track Tuning & Adjustment"* chapter for more information on chassis stuning with tire temperatures.

Checking Toe-Out/Toe-In

Toe is the difference in distance between the front and

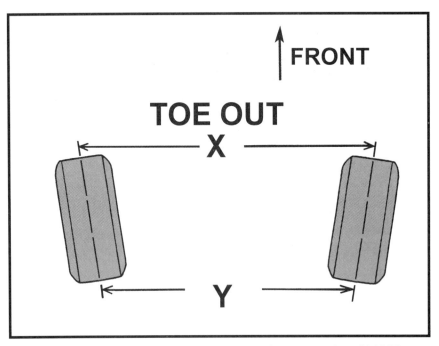

Toe-out is X minus Y, when the X distance is greater than Y. With toe-in, Y is greater than X. Quarter midgets do not use any amount of toe-in or toe-out, so be sure to check the car to see if any exists.

For working convenience, use a flat surface like a pit kart or large flat table, which can be leveled.

rear measurements of the front tires, measured on the center of the tread surface at spindle height. With toe-out, the front measurement is greater than the rear. Toe-in is the opposite. For different types of racing vehicles, toe-out or toe-in is used. However, using any amount of toe in or out will cause tire scrub as the car goes down the straightaways, because the tires are not pointing straight ahead. For a low-powered car like a quarter midget, this is not desirable. For that reason, set both front wheels straight ahead. Quarter midgets do not use any amount of toe-in or toe-out.

Be sure to check your car to see if any toe in or out is present between the front tires. To do this, measure the tread face of each front tire, and make a small mark at the center of the tread face of each. Then set some type of fixed pointer against the mark and spin the tire to make a slight scribe mark around the circumference of each tire. Next, measure between the two scribe marks at the front of the tires and at the back of the tires. Both measurements should be equal.

Get the Car Race-Ready

For the balance of the set-up procedure, the car has to be race ready and setting on a flat, level surface. For working convenience, this flat surface

Chassis Setup & Alignment

can be a pit kart or large flat table, both of which can be leveled to assure a flat surface. Once the car is assembled, it is important to check that the front wheels and tires do not rub or contact anything in the front suspension as the wheels are steered left and right. Steer the car to the left, lift the front wheels, and look for interference. Steer the car to the right and do the same.

Adding Weight for A Class

QMA class requirements provide for a minimum weight for the car and driver combined. Determine the minimum weight for your class, then weigh the car with the driver (including all safety equipment). Subtract the scaled weight from the minimum required weight, and you will know how much (if any) ballast you will have to add to the car. Be sure to add enough weight so that the car and driver are five to ten pounds over the required minimum weight just in case your scales differ from the club's official scales. To avoid having a problem with this, get weighed on the track's scales before or after practice. If you come up short on weight, add the extra ballast right away.

Small pieces of lead ballast can be bolted (securely!) to the chassis to increase the required weight. It is generally best to place the ballast centered in the chassis – usually right under or behind the driver. But if you need to increase the left side weight percentage or rear weight percentage of the car, bolt the ballast toward the left side or rear of the car. For pavement cars, place the added weight as low as possible in the chas-

The front wheels should be set in a close as possible, without any interference with the suspension. This is an ideal distance.

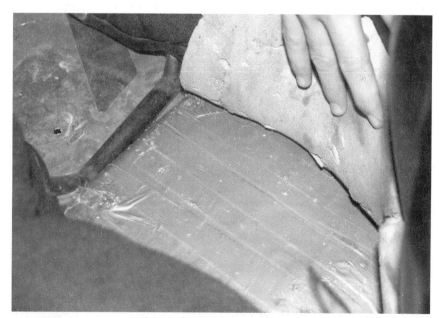
This is 30 pounds of lead ballast bolted below the driver's seat. This is an ideal location to add the ballast.

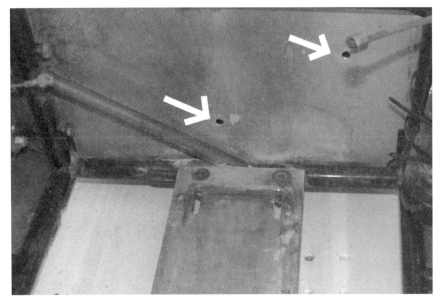

For cars running on a dirt track, the ballast can be placed higher on the chassis. It is generally mounted to the upper middle area of the seat back. Weight is also added in this area when the driver is shorter and lighter.

sis. For cars running on a dirt track, the ballast can be placed higher on the chassis. It is generally mounted to the upper middle area of the seat back. Weight is also added in this area when the driver is shorter and lighter so that there is adequate weight transfer to the right rear during cornering. This increases the overturning moment during cornering, which can help increase side grip for the right side tires.

If you find that you need to add a lot of weight to make a class weight, consider using a steel belly pan. It is much heavier than sheetmetal, and it places the weight as low as possible, and centered in the car. In most cases, a steel belly pan will add about 15 pounds.

Chain Tension

Once the rear axle is squared, proper chain tension must be set. The car should be in race-ready condition and setting on a flat surface, with the driver (or ballast equal to his weight) setting in the seat so the suspension is loaded.

Chain tension is the deflec-

Chain tension should be set at 0.5 to 0.75-inch up and down play on the chain.

Chain tension is set by moving the engine forward or rearward in the slots of the engine mounting plate.

 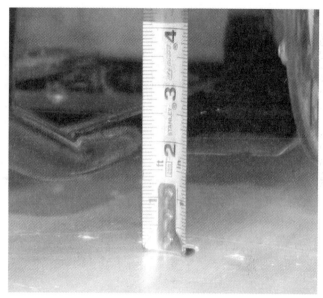

The ride height at each corner is measured from the bottom of the frame rail (where it meets the side panel) to the ground. Mark this place on the frame rail at each corner with a marker so you always measure to the same place. The ride heights at each corner affect the corner weights of the car, and how the chassis functions as the car corners.

tion in the chain measured half way between the front and rear gears. Use a long straight edge placed along the top of the chain to accurately measure the deflection.

Chain tension should be set at 0.5 to 0.75-inch up and down play on the chain. Chain tension being set too tight can bind the rear suspension movement of a car. Move the rear axle through full droop and rebound travel to make sure the chain is not too tight at those points or that it does not bind the suspension.

Chain tension is set by moving the engine forward or rearward in the slots of the engine mounting plate.

Chain deformation during cornering can be a problem with faster, heavier classes. These classes experience more chassis roll during cornering, so there is more angle change between the rear axle, where the rear sprocket is located, and the chassis, where the front sprocket is located. This angle change creates a twisting motion on the chain, which can cause accelerated wear, and even chain breakage.

One way to help minimize this chain-killing deformation is to mount the engine at an angle on the mounting plate. Place ¼-inch thick washers between the engine and the mounting plate on the right side. This makes the chain operate at a slight angle on the straightaways, but it helps to keep the chain operating angle less severe during cornering. It prevents the chain from operating at a severe angle during cornering which kills the life of the chain.

Setting Ride Heights

Before the ride heights are set, make sure the car is completely race-ready. The tires should be inflated to the proper racing pressures and the fuel tank should be half full. The driver with full safety equipment (or ballast equivalent) must be in the driver's seat, and any required ballast must be in place. Set the car on a flat level surface so that floor or surface irregularities will not influence the settings. Be sure that the front tires are pointing straight ahead (because steering the front wheels to one side or the other

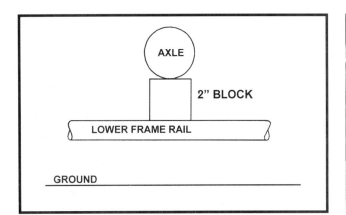

Once the baseline setup established, a quick way to reset ride heights and corner weights is by using chassis blocks. Ride height blocks, made from wood, are set between the axles and the lower frame rails.

After the correct ride height is set at each corner of the car, carefully measure between the axle and the lower frame rail, and cut a piece of 2x4 to this dimension.

will change corner heights due to spindle kingpin angle and caster).

Set the ride height on the left side first, then the right side. Then recheck the left side again. Setting the heights on one side may change the heights on the other.

The ride height at each corner is measured from the bottom of the frame rail (where it meets the side panel) to the ground. Mark this place on the frame rail at each corner with a marker so you always measure to the same place. The ride heights at each corner affect the corner weights of the car, and how the chassis functions as the car corners.

Chassis builders design their cars to operate most efficiently at certain corner heights. Be sure to obtain these specifications from your car builder. The basic idea of setting the four corner heights is to achieve a slight forward tilt and a slight left side tilt.

To set ride height, adjust the collar on each coil-over to raise or lower the ride height. Turning the collar clockwise will raise the corner, and turning it counter-clockwise will lower that corner. Most shocks are threaded at 8 threads per inch, so making one full turn on the shock collar will change the ride height at the shock by 1/8-inch.

If you don't have specific ride height numbers from your car builder to start with, use these:

Asphalt Track
LF 1"   RF 1 1/4"
LR 1 1/4" RR 1 1/2"

Dirt Track
(Packed dirt, good grip)
LF 1 1/2" RF 1 1/4"
LR 1 1/4" RR 1 1/2"

Dirt Track (Slick)
LF 1 3/4" RF 1 3/4"
LR 1 3/4" RR 1 5/8"

Ride Height Setting Shortcut

Once you have the baseline setup established, a quick way to reset ride heights and corner weights is by using chassis blocks. This method of setting ride heights has been used for years with sprint cars and midgets. Ride height blocks, made from wood, are set between the axles and the lower frame rails. After the correct ride height is set at each corner of the car, carefully measure between the axle and the lower frame rail, and cut a piece of 2x4 to this dimension. The block should be able to slip between the axle and frame rail with a tight fit. Be sure to plainly mark each block with paint or a Sharpie marker which corner of the car the block fits, and also mark on the block the correct height of the block.

To use the blocking method of setting ride height, set the car on a flat level surface. Place the proper block between the axle and frame rail at each corner of the car. Loosen the adjusting collar on each coil-over to take any load off the springs. The suspension will be supported on the blocks. Working with one corner at a time, tighten the adjusting nut against the coil-over spring until it is seated snugly against it. Don't tighten it enough to compress the spring. Then tighten the set screw on the adjusting collar, and use tape around the threads of the coil-over body so that the adjuster does not back off. Remove the blocks and you are ready to race.

After using the blocks for a while, be sure to check their height periodically because they may get shorter due to wear and use. That is the reason that their correct height should be marked on the block.

Scaling the Chassis

Scaling the car is the process of setting the individual wheel weights to achieve the desired rear weight percentage, left weight percentage, and cross weight percentage. This is a critical setup step, and you cannot skip it and expect to have a competitive race car.

Scaling the car is the process of setting the individual wheel weights to achieve the desired rear weight percentage, left weight percentage, and cross weight percentage. This is a critical setup step, and you cannot skip it and expect to have a competitive race car.

(Right) Bathroom scales work great for scaling a quarter midget, but make sure the scales are accurate and repeatable.

Setting the ride height at all four corners according to the car builder's specifications will get you in the ballpark for corner weights. But using wheel scales is the only way to accurately determine the chassis' weight percentages.

There are several types of wheel scales that can be used for scaling the car. Many racers use household bathroom scales. There are cheap bathroom scales, and there are more expensive bathroom scales. Sometimes price does not yield quality. What you have to look for is accuracy and repeatability. To test a scale for these qualities, use a

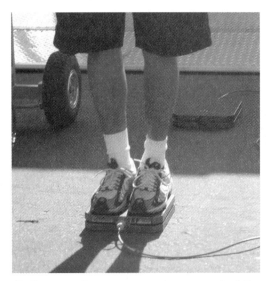

Before scaling a car, make sure all of the scales are showing the same weight. Test each scale with a known amount of weight. Each scale should register the same.

An excellent type of purpose-built scale is Longacre Racing's Computerscales. They offer a 1/10 percent accuracy and 400-pound capacity. Electronic scales do all the math for you.

known quantity of weight, such as a 50-pound piece of ballast. Place the weight on the scale and see what the scale says it weighs. If it isn't 50 pounds, the scale isn't accurate. Place the weight on the scale time and time again. Each time the scale should register 50 pounds. If it doesn't, the scale is not repeatable and you shouldn't use it.

Using bathroom scales is better than nothing, but if you want to be accurate and repeatable, you need to use electronic wheel scales made for the specific purpose of weighing a race car. An excellent type of purpose-built scale is Longacre Racing's Computerscales. They offer a 1/10 percent accuracy and 400-pound capacity.

When the car is scaled, it should be race-ready, but without the driver. The driver is not included for scaling because the driver is a big variable to the true corner weights. If the driver is sitting upright in the car, the weight distribution is one thing. But if the driver leans to the left, it changes all the corner weights. The car is scaled without the driver for the baseline setup, and then the car is adjusted at the track. Be sure to bolt on any additional ballast you need to make class weight.

When using wheel scales, it is extremely important that all four scales are set level to each other. If they are not, the differences in height are going to add or subtract cross weight, or left or rear weight in the chassis. Check that the floor is level front left to right, front to rear, and diagonally. Use a line level to check for accuracy. Just eyeballing it won't do. If a scale needs to be shimmed up, an ideal shim material is a 12x12 linoleum square, or 12x12 pieces of sheetmetal. This works great if you are setting up the scales on the ground in the pits. In the shop, when setting the car on a pit cart or table to scale it, make sure there is a means of adjustment for leveling. Many racers use a table for setup and scaling which has screw-in/screw-out leveling pads at the end of each table leg.

Before electronic scales are used, they should be turned on (without any weight on them) and zeroed out. Read the instructions that came with the scales to learn how to do this. If you do not zero the scales first, you will get faulty weight readings.

To help you get your chassis set up properly, wheel scales must be used in the same manner every time, or you will not get repeatable results. When scaling the car, always make sure:

1) The car weighs the same each time (unless you purposely added or removed ballast);

2) The fuel load is always the same;

3) The car is bounced up and down several times to eliminate any binding in shocks and springs;

4) Tire pressures are always the same;

5) Tire stagger is always the same (unless you purposely intend to change your stagger to change your setup);

6) Each wheel scale is always centered under the center of each tire.

The next step is to raise the car up in place and put a wheel scale under each wheel. Once the car is in place, bounce the car up and down to settle the suspension.

When scaling the car, you are looking for a certain percentage of the total car weight (without the driver) as left side weight, rear weight and cross weight. Left side weight percentage is the total of the left front and left rear corner weights divided by the total car weight times 100 to make it a percentage. Rear weight percentage is the total of the left rear and right rear corner weights divided by the total car weight. Cross weight percentage is the total of the right front and left rear corner weights divided by the total car weight.

The target percentages for a paved track car are:
Left   56 to 59%
Rear   60 to 62%
Cross 47 to 52%

The left rear corner weight should be 12 to 15 pounds heavier than the right rear corner weight.

Be sure that all four scales are set level to each other. If they are not, the differences in height are going to add or subtract corner weight.

Our sample car for scaling is running in a 260-pound class, and the track being run has medium banking with wide turns, and is smooth asphalt. The goals for the weight percentages are:

57 to 58% left
61% rear
51% cross
Left rear 12 to 15 pounds heavier than the right rear.

Read the wheel weights of your car and record them on a grid pattern on your paper for ease of visualizing the corners. The weights shown on the wheel scales are:
LF 40   RF 32
LR 65   RR 47

Chassis Setup & Alignment

To adjust the car for the desired corner weights, make a small change at all four corners of the car. Making a small change at each corner will allow you to maintain proper ride height and still move the required amount of static weight.

This is 184 pounds total weight with ballast on the car. The driver weighs 80 pounds, so the total class weight is 264 pounds, including a 4-pound safety margin.

If you are using electronic wheel scales, the left, rear, and cross weight percentages will be calculated for you. But if you are using digital bathroom scales, here is how the percentages are calculated:

First, add all four corner weights together to obtain the total car weight:

    40
    65
    32
    47
184 total car weight

Now calculate the percentages:

Left Side
    40 + 65 = 105
    105 / 184 x 100 = 57%
Rear
    65 + 47 = 112
    112 / 184 x 100 = 61%
Cross
    32 + 65 = 97
    97 / 184 x 100 = 53%
Left rear corner weight
    65 − 47 = 18 (LR heavier than RR)

Looking at the percentages, the left side percentage and rear percentage are very close to the goals. But the cross weight percentage and left rear corner weight are too high. How is this adjusted to meet the goals?

To adjust corner weights and cross weight, turn the adjusting collar on each coil-over up or down. To jack weight in (add weight to a corner), or raise a corner height, turn the adjusting collar down (clockwise). To take weight out, or to lower a corner height, turn the adjusting collar up (counter-clockwise).

To minimize making changes in the chassis ride height, an adjustment is made at all four corners of the car. The adjustment could be made at just one corner, but that requires a large change in ride height at just one corner. That is not acceptable. Making a small change at each corner will allow you to maintain the proper ride height at each corner and still move the required amount of static weight.

To adjust this sample car, make one turn in (or down) on

the left front and right rear adjusting collars, and take one turn out (or up) on the right front and left rear adjusting collars. To visualize this:

LF 1 turn in   RF 1 turn out
LR 1 turn out   RR 1 turn in

The new corner weights are:

LF 42   RF 30
LR 63   RR 49

The new percentages are:

Left
105 / 184 x 100 = 57%

Rear
112 / 184 x 100 = 61%

Cross
93 / 184 x 100 = 51%

Left rear corner weight
63 – 49 = 14 pounds (LR heavier than RR)

Comparing these new numbers to the initial numbers, we find that the left side percentage and rear percentage did not change, even though the individual corner weights did change. The cross weight remained the same even though the corner weights changed. The left rear corner weight differential is lower by 4 pounds to 14 pounds. The adjustments in corner heights transferred 2 pounds from the left rear to the right rear, and shifted 2 pounds from the right front to the left front. Doing this will change the diagonal percentages, but never the front-to-rear percentages and left-to-right percentages.

If you need to change the left side weight percentage, or rear percentage, the only way to accomplish this is by physically moving weight around on the chassis. Usually this is done by moving ballast weight attached to the chassis. For example, if more left side weight is required, move the ballast toward the left wheels. If more rear weight is required, move the ballast toward the rear.

Be sure to keep records of your ride heights and corner weights in a notebook so you know what your starting baseline setup was. Always use a chassis baseline setup sheet so you know all of your chassis settings. Keep good notes on everything you do!

Once the adjusting collar is set on the coil-over for the correct weight and height, it must be anchored in place so that vibration does not cause the collar to back off. This collar uses a set screw to anchor it in place. As a backup, the threads above the collar are taped to prevent the collar from backing up. It was also very smart to mark which corner of the car the shock belongs on.

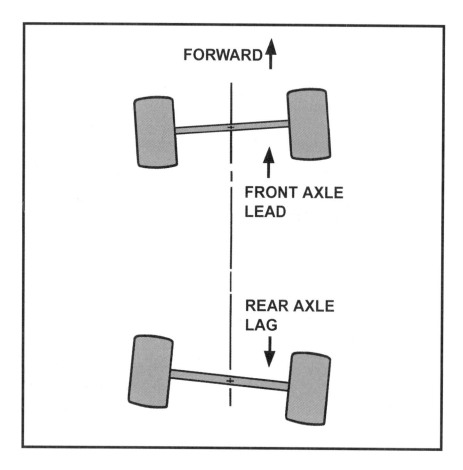

Scaling the Car After A Race

Scaling the car on a regular basis gives you feedback about chassis performance. After you get your race car back to the shop and cleaned up, you want to scale the car so you know how changes you made at the track affected the baseline setup. This can tell you a lot about the handling performance of your car. In your notes, you should have the setup you started with. You should also have notes on every chassis setup change you made at the race track, along with comments on how these changes affected the handling. So now when you scale the car and measure ride heights, and compare this data to your notes, you will have a good understanding of how the changes you made at the track affected the handling performance of the car. Compare the corner weights and heights to the baseline setup you started with. Did the changes make the car faster? Did the changes make the driver feel more comfortable? If so, you may want to incorporate the changes into your baseline setup.

Front Axle Lead

Quarter midgets typically race on very short tracks. With short straightaways, that means the cars spend most of their time in the corners turning left. To help a car turn left more easily, the right front wheel can be set ahead slightly. That is called right front axle lead. Pushing the right front tire ahead of the left front makes the car tend to turn left naturally.

Typically, the right front is set ahead of the left front 1/8-inch to 3/16-inch. Leading the right front helps the car turn in to the corners easier. The correct amount of axle lead is a fine line compromise between too much lead, which makes the car pull to the inside of the track on the straightaways, and too little axle lead, which makes the car want to drive forward rather than turn into a corner.

The front and rear axles should already have been squared to the chassis, so the measurement between the front and rear axles on both sides should be the same. Double check this before you make any changes. The lead is adjusted by shortening the left front radius rod. Disconnect the rod end from the axle, turn in the rod end one to two turns, and reconnect to the axle. Take a measurement again. For 1/8-inch lead, the left side wheelbase measurement should be 1/8-inch less than the right side wheelbase. Reassemble the car and make sure that nothing rubs between

Chassis Setup & Alignment

the wheel, suspension and chassis.

Be sure to test the axle lead with your driver to make sure it feels comfortable. Too much axle lead may make the car feel like it pulls too much to the left on the straights and at turn entry. On the other hand, if axle lead isn't used, the driver may complain that the car is too hard to turn into the corners. This is an adjustment that has to be tested with the driver.

Rear Axle Lag

For the same reason presented in "Front Axle Lead," rear axle lag can help the race car turn left more easily. Rear axle lag sets the right rear wheel behind the left rear slightly. This arrangement makes the rear of the car steer outward in the turns, which makes the car freer and turn left easier. Typically, the right rear would be set behind the left rear 1/8-inch. This is set by adjusting the upper and lower radius rods on the right rear corner 1/8-inch longer. Make equal adjustments on the upper and lower rods. This adjustment is made after the rear axle is squared and the rear birdcages are leveled. Double check that adding axle lag did not change the birdcage timing.

Rear axle lag is an advanced chassis tuning tool. Use this only if the car has a difficult time turning through the middle of the corner, after you have exhausted all other chassis tuning adjustments.

Choosing Spring Rates

Spring rates are chosen to coincide with or adapt to track conditions, track size and track speed, as well as class weight and driver size. Spring rates will vary depending on track configuration. For example, a track with tight short radius turns will require different spring rates than a track with longer sweeping radius turns. And, on dirt, heavy or tacky tracks will require different spring rates at each corner of the car than a dry slick track.

There is a problem in trying to make specific spring rate recommendations for all quarter midget cars because there are so many variables among the different chassis manufacturers. Manufacturers build their cars and suspension linkages differently, and the cars can have a tendency to be either loose or tight at corner entry or corner exit, so then they design in other factors in the suspension to deal with

Spring rates are chosen to coincide with or adapt to track conditions, track configuration and speed, as well as class weight and driver size. Start with your chassis manufacturer's recommendation, then test and adjust from there. Many times you will find that two drivers with the same brand of car on the same track will use different rate springs.

When a spring is mounted at an angle away from true vertical, it loses rate. Variable mounting positions for coil-overs can be used to tailor the spring rate slightly. By moving from the outside to the inside mounting position, the rate gets slightly softer.

these tendencies.

With coil-over suspensions, there are three major factors that affect the wheel rate of a spring (which is the effective spring rate at the wheel): how close the spring is mounted to the wheel, the mounting angle of the spring from true vertical, and how far below the axle centerline the spring is mounted. These three items do vary among chassis manufacturers.

When a spring is mounted closer to a wheel, the rate can be softer than if it was mounted further away from the wheel, because there is less leverage acting on the spring. When a spring is mounted at an angle away from true vertical, it loses rate. For example, a spring mounted 10 degrees away from vertical has a 3 percent spring rate loss. A spring mounted 20 degrees away from vertical has a 12 percent spring rate loss. A coil spring mounted lower below the rear axle centerline has to be slightly stiffer in rate than a spring mounted higher on the axle. That is because the lower mounted spring receives a greater weight load transfer under acceleration than a higher mounted spring.

The bottom line is that there are way too many variables among the different makes of chassis, and race track applications, to be able to specify spring rate requirements that will work on all chassis brands. There is not a uniformity in suspension design among chassis builders, even though many chassis appear similar. Check with your chassis builder for their baseline spring rates for your type of track. Then use the information we provide to you in the next chapter – *"Track Tuning & Adjustment"* – to adjust and improve your setup, and adapt to changing track conditions.

Keep in mind that the baseline spring rates are a starting point. The final spring rate selection for a car depends on many factors including driver weight and height, class, track surface, track banking and configuration, car balance and driver feel.

Track Tuning & Adjustment

Chapter 6

Suspension and handling is the most critical element in going fast in a quarter midget. It is a combination of a lot of things – handling balance, chassis geometry, proper tire loading during cornering, and a reduction of tire scrub. And what is really important is understanding how to adjust the chassis to maximize these items and accommodate changing track conditions.

Chassis Setup Philosophy

Novice drivers should have a tight car. It should be tight enough to keep the rear end from drifting out during cornering to stop it from being loose. On a loose car, the rear end slides. It may only be a slight amount of sliding, but a novice may feel unsure and uncomfortable. A tight car will allow a novice to concentrate on his driving line and build confidence. Once a driver has learned his driving line and gained confidence, you can loosen up the car in small increments by moving the right rear wheel out a small amount.

Novice drivers should have a tight car. It should be tight enough to keep the rear end from drifting out during cornering to stop it from being loose. This will make the driver more comfortable. Once a driver has learned his driving line and gained confidence, loosen up the car in small increments.

Watch the Driving Line

Watch your driver. Make sure he is driving the proper line. If he isn't driving the proper line, you cannot make meaningful chassis adjustments. For example, if the driver turns in too early for a corner and heads to the bottom, by the time he gets to mid-corner the car is going to push as he makes a hard left turn through the middle. In this instance, the car isn't pushing, the driver is driving the wrong line. He is forcing the car to make too tight of a turn. Spend some time orienting the driving to get him on the proper driving line. Once he can run the proper line lap after lap, feedback about push or loose conditions will

Watch the driver carefully. If he isn't driving the proper line, you cannot make meaningful handling adjustments. Once the driver is on the proper line lap after lap, feedback about push or loose conditions will become meaningful.

Work with the driver to minimize hand movement. Smooth, calm hands produce fast laps. If the driver corrects and overcorrects, he is going to upset the chassis.

become meaningful.

Work with your driver to minimize hand movement. Smooth, calm hands produce fast laps. If the driver corrects and overcorrects, he is going to upset the chassis. It will rock back and forth and be unstable, loading and unloading the tire contact patches. This will cost the driver valuable tenths of a second. To drive the proper line and be fast, a minimum of hand movement is required. The driver has to have confidence in his setup and not feel that he has to make a correction for every little movement that the car makes.

All chassis adjustments discussed in this chapter assume that the car is being driven on the proper driving line.

Chassis Sorting at the Track

The car is set up properly at the shop. The springs and shocks should be correct. And yet, once the car hits the track, there may be several handling problems to sort out. In this section, we will take you through these problems, isolating particular handling problems which are common to quarter midget racers, and offer a diagnosis and correction for each.

All quarter midget race cars will react to chassis adjustments in the same manner regardless of which class a car races in. The differences in classes are in total weight, spring rates and horsepower. But the chassis adjustments work the same for all classes.

When sorting the chassis at the race track, only one adjustment should be made at a time. If more than one change is made, you will never be able to determine the effect of just one change on the handling.

Always keep a notebook with you at the race track. Besides recording all the usual starting specifications of the car, you should also record each handling problem, and what adjustments were made to correct the problem, and how these changes affected the handling. Also, keep good notes on weather conditions and how the track changed during the day, and how you adjusted the chassis for this. This log of information is important to you for two reasons: 1) Should the changes made not solve the problem, or should they make things worse, you know what changes not to make, and how to get back to your original setup. 2) It gives you a complete log of experience that tells you what changes on the car are effective for specific handling problems.

What Adjustments Are Available and What Do They Do?

What are the various chassis tuning adjustments available, and what do they do to alter the handling characteristics of the car? There are a wide variety of adjustments available to the racer to tune the chassis to track conditions, or to correct

All chassis adjustments discussed in this chapter assume that the car is being driven on the proper driving line.

handling problems. Some are easier to make than others. Some have a very quick and direct influence over handling, while some have a more subtle influence. Some adjustments will influence the chassis more at one phase of cornering, while others will have an effect through all three phases of cornering (which are corner entry, mid corner and corner exit). Every change that can be made to the chassis to alter handling characteristics is listed below, and following the list is a discussion of how each adjustment changes the chassis. This list includes:

Stagger
Wheel spacing/tracking
Tire pressure
Front Panhard bar height
Rear Panhard bar height
Front axle lead
Chassis tilt

Rake
Cross weight
Shock absorber valving
Spring rates
Tire compounds

Stagger

Stagger is the difference in inches of tire circumference between the left rear and right rear tires. Measure the circumference of each tire – mounted on the proper wheel – with the air pressure set to the desired hot pressure for the tire. Then subtract the left rear circumference from the right rear circumference. The difference is the amount of stagger. Use the hot inflation pressure for the measurements because that is what the tire experiences on the track.

In most track applications, an unlocked car will use 2 inches to 3 inches of stagger. On an unlocked car, stagger is

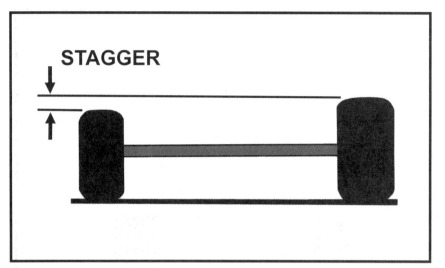

Stagger is the difference in inches of tire circumference between the left rear and right rear tires.

Track configuration and track conditions dictate the amount of stagger used. Stagger affects the corner weights and corner heights of the car. Using more stagger will help the car to turn through the middle and at corner exit.

used to change ride height and corner weight settings.

If your car's left rear is locked, be careful that you don't have too much rear stagger. That will cause tire scrub down the straightaways and kill straightaway speed. Four inches should be considered a maximum. If you find that your car requires more than four inches of stagger to turn through a corner, another type of change in the chassis may be required. Try using a right rear spring that is one step stiffer in rate.

When using a locked axle or a left rear ratchet hub, stagger is required because two tires on opposite ends of a solid axle are running on two different radii through a turn. The outside tire must travel a further distance on a wider arc than the inside one. This is accomplished by the outside tire being larger in circumference than the inside tire so it runs at a slightly faster speed. The tighter the radius of a corner, the larger the circumference of the outside tire must be in relationship to the inside. So, the tighter the corner radius, the more stagger required.

Track conditions also dictate the amount of stagger used. If you are having a hard time turning the race car, stagger can help accomplish it. It will make the right rear overdrive the left rear and drive the car in a tighter arc – it makes the car turn itself.

Stagger changes the corner weights of the car. That is because a larger circumference tire is taller, so it raises the corner height of the car. When a larger right rear tire is used, it adds weight to that corner and puts more tilt into the chassis. When a larger left rear tire is used, the corner height at that corner is raised and more weight is added to that corner. It also takes out chassis tilt and takes weight off the right rear and adds it to the left rear.

Stagger puts negative camber into the right rear tire, and positive camber into the left rear tire. This will put more weight on the inside edge of the left rear and right rear tires. (Inside edge is the edge toward the infield.) One way to tell if the car has too much stagger is to read the tire wear pattern. If there is a lot of heat on the inside edge of the right rear as compared to the rest of the tire, then there is too much stagger. The slight amount of negative camber that the stagger produces is helpful because it counteracts tire deformation during cornering, which flattens out the tire contact patch on the track.

Stagger is effective under acceleration when one wheel is driving around the other when using a locked rear axle or with a ratchet hub. More stagger will make the right rear drive around the left rear. Less stagger makes the left rear more dominant.

Wheel Spacing or Tracking

Wheel spacing, or tracking, is the process of moving the wheels in or out on an axle in relationship to the centerline of the race car. Rear wheels can be moved with axle spacers. Widening or narrowing the rear track width of the car by spacing the right rear in or out changes weight loading on the right rear tire during cornering. This is one of the most-used handling adjustments

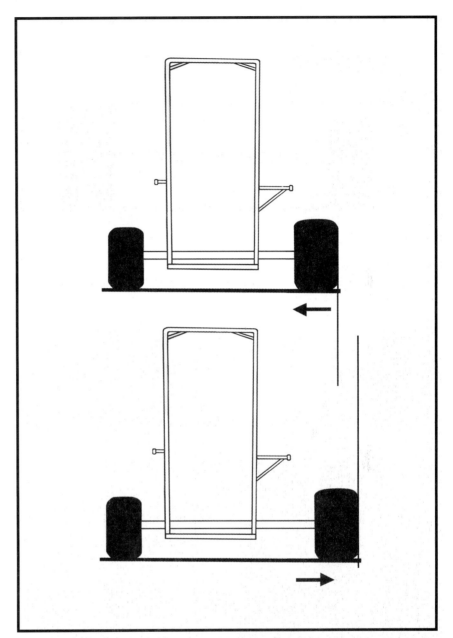

Wheel spacing, or tracking, is the process of moving the wheels in or out on an axle in relationship to the centerline of the race car. Widening or narrowing the rear track width of the car by spacing the right rear in or out changes weight loading on the right rear tire during cornering. This is one of the most-used handling adjustments there is for a quarter midget.

Although the right rear is almost always used for wheel tracking adjustments, having the axle protrude past the left rear wheel an extra 1/2-inch provides the opportunity to space the left rear out slightly.

To determine the wheel tracking – or tire offset – measure from the outside of the frame rail to the center of the right rear tire.

inward closer to the chassis increases weight loading on it during cornering and makes the car tighter.

In almost all situations, the left rear tire is set as close to the chassis as possible. However, experienced racers will set their rear axle protruding with an extra ½-inch through the left wheel to give them the ability to space the left rear out slightly. Moving the left rear out away from the car will tighten the chassis. Moving it in this direction increases right rear weight. Moving the left rear outward functions the same as moving the right rear inward more when there isn't any more room left to move it in.

To determine the wheel tracking – or tire offset – measure from the outside of the frame rail to the center of the right rear tire. This distance can be adjusted by changing rear axle spacer widths. A normal baseline tire offset for most quarter midgets is 10 to 12 inches. The wheel can be moved in or out to adjust for track conditions from that point. Check with your chassis manufacturer for a specific baseline starting point for your car. The exact distance for your race car depends on many factors, including class, overall weight, driver weight and height, track surface, track banking, tire compound, and driver feel. There are many factors involved, and the ulti-

there is for a quarter midget.

Moving the right rear wheel outward from the chassis centerline decreases weight loading on it, which makes the car looser. Moving the right rear mate distance is just part of the total handling package.

The measurement from the chassis to the center of the right rear tire should always be a step in the baseline chassis setup. Always record this distance in your setup notes. Because this is such an important tuning tool, always keep notes of how you have changed this distance at the track, why, and how the changes affected the handling.

A note on the baseline setup: if the driver is taller and in a heavy class, it would be best to space the right rear tire out an additional 1/4 to 1/2-inch. This is because the additional overturning moment caused by the driver's extra weight and height is transferred to the right rear tire which tightens up the car. If the right rear tire is moved outward slightly, it will loosen the chassis and compensate for the extra weight transfer.

In almost all situations, the front tire offsets are not changed. The chassis is most easily adjusted by spacing the right rear wheel in and out. However, in some extreme cases, the right front can be spaced in or out using wheel spacers. The effect of this is the same as at the rear axle – it will lessen right front tire grip and side bite. This right front offset change is made only in cases where the right front sticks so well that it is impossible to adjust the chassis using other means.

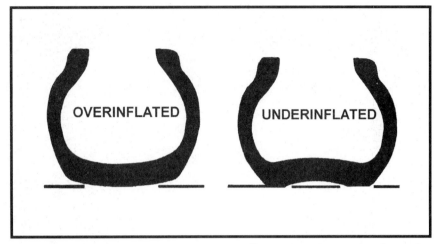

Overinflation causes the tire to be stiffer and reduces tire footprint on the track. Underinflation causes the tire to flex more, and reduces center tread contact when the underinflation is severe.

When using lower right side tire pressures, always use bead lock wheels to make sure the tires stay seated on the wheels. Low inflation pressure can cause a tire to roll off the rim.

Tire Pressure

Tire pressure affects the stability of the race car and the tires' grip on the track. Adding or subtracting tire pressure can be a quick fix to adjust handling. If the car has a push, add a pound or two of air

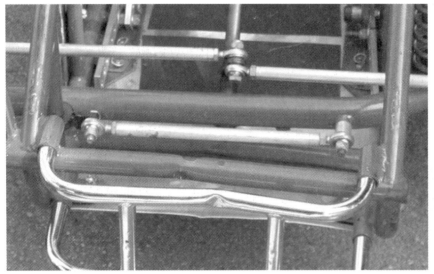

The height of the front Panhard bar establishes the front roll center height. Start with the bar at a suggested height established during the baseline setup. In general, the front Panhard bar adjustment is rarely used to make handling adjustments.

pressure in the rear tires to take away some rear grip. If the car is loose, reduce rear air pressure by a pound or two to increase rear grip.

Increasing tire pressure takes away grip. Decreasing tire pressure increases grip.

Normal tire pressures on pavement range from 15 PSI to 20 PSI on the right side. Sometimes lower right side tire pressures are required on flatter tracks or slippery tracks. When using lower right side tire pressures, always use bead lock wheels to make sure the tires stay seated on the wheels. Low inflation pressure can cause a tire to roll off the rim. On dirt tracks, always use bead lock wheels on the right side.

Adjusting Tire Pressures

Tires will build up more air pressure the first time that they are heated up from a cold tire to race temperature. When you bring the car in to the pits after the first run on new tires, you will see an increase in pressures from 2 PSI to 3 PSI. Reset the pressures to the cold starting pressures, and send the car back out on the track while the tires are still hot. You will find that the pressures will build up less this time. Be sure to keep good records. This will give you good feedback on the starting pressures you should use.

Front Panhard Bar Height

The height of the front Panhard bar establishes the front roll center height. Start with the bar at a suggested height established during the baseline setup. Lowering the bar from that point lowers the front roll center. This increases weight transfer across the front and adds more grip at the right front tire. Raising the bar above the baseline setup raises the front roll center, reduces weight transfer across the front, and reduces tire grip at the right front.

If the right front washes out at corner entry and won't grip the track, lower the front Panhard bar. This will produce more grip and side bite.

In general, the front Panhard bar adjustment is rarely used to make handling adjustments. It is only used when there is a problem with the right front tire not having enough grip, or it has an extreme amount of grip, and no other adjustment will work.

Rear Panhard Bar

Raising or lowering the rear Panhard bar raises or lowers the rear roll center in a manner as described above with the front Panhard bar. Adjusting the rear Panhard bar height is a quick and easy adjustment. Lowering the bar and roll center will tighten the chassis. Raising the bar and roll center will loosen the chassis. This adjustment has the most effect on the chassis at midcorner and corner exit.

The rear Panhard bar can be

adjusted on the left side, on the right side, or both. The quickest and easiest adjustment is made on the right (chassis) side.

Be aware that adjusting the bar on just one side makes an incremental change in height that is different than if it is adjusted at both ends. If the bar is raised ½-inch at both ends, the roll center (which is located at the center of the bar) is raised ½-inch. But if the bar is raised ½-inch on just the chassis side, the center of the bar is raised much less than ½-inch because the bar is pivoting about its fixed mounting point on the left side.

A quarter midget chassis is very sensitive to rear Panhard bar height. Make adjustments in ½-inch increments. Lowering the bar will tighten up the chassis, so be very careful. If you get the bar too low it will stick the car too tight and cause it to bicycle (lift both left wheels off the ground). This is a very dangerous situation because the car can tip over during cornering.

Front Axle Lead

Front axle lead is used to help a car turn left more easily. Axle lead sets the right front wheel ahead of the left front slightly. Leading the right front tire ahead of the left front makes the car tend to turn left naturally.

Typically, the right front is set ahead of the left front 1/8-inch

Raising or lowering the rear Panhard bar raises or lowers the rear roll center. Adjusting the Panhard bar height is a quick and easy adjustment, but that adjustment is only used when other options are more time consuming to make.

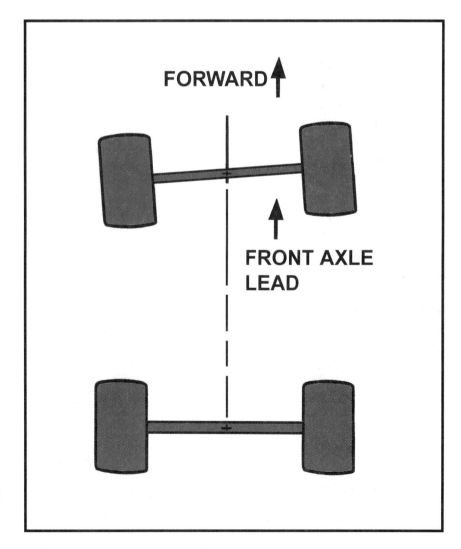

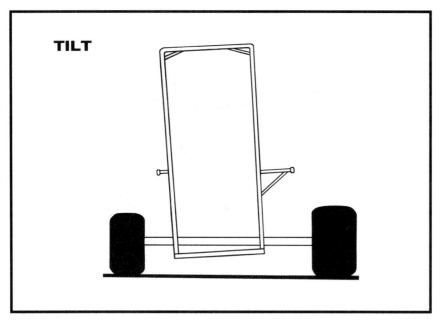

Tilt is the attitude of the chassis, as the chassis is viewed from the rear, in relation to the flat level ground. Tilt increases the angle of the chassis (right side higher than the left side).

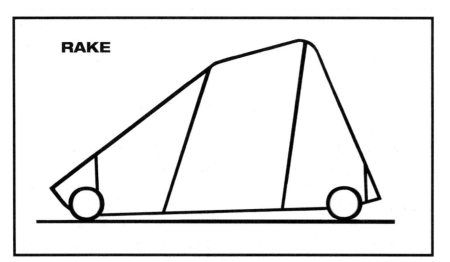

Rake is the attitude of the chassis when viewed from the side. Setting the rear ride height of the chassis higher than the front is referred to as rake.

to 3/16-inch. Leading the right front helps the car turn in to the corners easier. The correct amount of axle lead is a fine line compromise between too much lead, which makes the car pull to the inside of the track on the straightaways, and too little axle lead, which makes the car want to drive forward rather than turn into a corner.

Be sure to test the axle lead with your driver to make sure it feels comfortable. Too much front axle lead may make the car feel like it pulls too much to the left on the straights and at turn entry. On the other hand, if axle lead isn't used, the driver may complain that the car is too hard to turn into the corners. This is an adjustment that has to be tested with the driver.

Chassis Tilt

Tilt is the attitude of the chassis, as the chassis is viewed from the rear, in relation to the flat level ground. Tilt increases the angle of the chassis (right side higher than the left side). Adding chassis tilt adds weight to the right rear and takes weight off the left rear. Rear ride heights will vary according to the tire circumference used. Using a larger right rear tire or increasing stagger adds more tilt.

Chassis tilt can be changed by making small changes on both of the left side coil-over adjusters to adjust the chassis height. To add tilt without affecting the chassis weight setting, take out 1/2 of a turn on the left rear and left front adjusters.

More tilt loosens up a chassis. Less tilt tightens a chassis. In some situations, such as on a flat slippery track, it is necessary to use reverse tilt, which is a situation where the right rear corner is lower than the left rear corner of the car.

Rake

Rake is the attitude of the

chassis when viewed from the side. Setting the rear ride height of the chassis higher than the front is referred to as rake. Setting the front chassis ride height higher than the rear is called reverse rake.

Raising or lowering the entire front or rear of the car is an effective method of adjusting for a loose or pushing condition, whether the rear axle is locked or unlocked. Lowering the entire rear of the car (or raising the front of the car) will help correct a loose condition. Raising the entire rear of the car (or lowering the front of the car) will help to correct a pushing condition.

Cross Weight

Cross weight is the combination of left rear and right front corner weights. Adding weight at the right front will also load the left rear and make it heavier, and adding weight at the left rear will increase right front weight.

Left rear weight means the left rear corner of a car is more heavily weighted than the right rear corner. With a locked axle or ratchet hub, left rear weight tightens up the car by making it drive in more of a straight line by favoring the left rear tire.

Shock Absorber Valving

Shock absorbers control the rate of weight transfer during cornering, they control spring movement, and they control suspension movement over bumps and surface undulations. Being able to control the chassis with the proper shock absorbers is a key element for proper handling. Different rates of shock valving and split valving shocks can be used to help control handling problems or to induce desirable handling characteristics. Refer to the "Shock Absorbers" chapter for complete information on tuning the chassis with shocks.

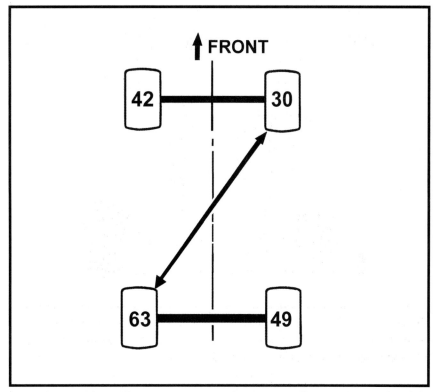

Cross weight is the combination of left rear and right front corner weights. Adding weight at the right front will also load the left rear and make it heavier, and adding weight at the left rear will increase right front weight.

Spring Rates

Spring rates control dynamic weight transfer during cornering. Springs provide roll stiffness or roll resistance against the weight being transferred. The stiffer the spring rate of a spring, the more roll stiffness or resistance it provides.

The front-to-rear handling characteristics of a car can be tailored by adjusting the front-to-rear roll stiffness proportioning. This is called roll couple distribution. It is how the cornering weight transfer is distributed between the front suspension and rear suspension. The end of the car with the highest amount of roll stiffness will receive the greatest amount of weight transfer during body roll.

The tire compound can be found on the tire sidewall. An SL4 compound is a harder compound and produces reduced track grip.

The resistance created by the front and rear suspension systems can be adjusted with spring rates to change the weight transfer distribution between the front and rear. For example, increasing a right front spring rate increases the front roll stiffness of the car. It makes more of the cornering weight transfer occur at the front of the car, and makes the car tend toward understeer, or push. Using a softer right rear spring will produce the same result because it increases the front roll stiffness compared to the rear.

Increasing a right rear spring rate increases the rear roll stiffness of a car, which shifts a greater proportion of cornering weight transfer to the rear. This makes the car tend toward oversteer, or being loose. Using a softer right front spring rate will produce the same result.

In most cases, if the proper springs are used on the car, handling changes can be effected on the chassis with the various chassis tuning elements (jacking weight at the corners, wheel tracking, Panhard bar height, etc.). But when these adjustments are ineffective, spring rates have to be changed. For example, if the car has a severe push at corner entry and tuning adjustments will not cure it, change the right front spring to a slightly softer rate. Or, if the car is consistently loose, and wheel tracking and weight jacking do not cure the problem, change to a softer spring rate at the right rear.

In summary, using a stiffer right front spring or a softer right rear spring makes the chassis tighter. Using a stiffer right rear spring or a softer right front spring makes the chassis looser.

Tire Compounds

Using different tire compounds on the front and rear wheels can be a very effective way of tuning the chassis for pushing or loose conditions.

If a car is pushing, using a harder compound tire on the right rear reduces rear tire grip and loosens the chassis. Or, using a softer compound on the right front will increase front tire grip.

If a car is loose, using a softer compound tire on the right rear will increase rear tire grip and tighten the chassis. Or, using a harder compound on the right front reduces front tire grip.

Toe Out

Quarter midget race cars do not use any amount of toe-out or toe-in. Be sure that you have checked this when doing the baseline chassis setup.

If the race car has banged the front wheels with another car on the track, check for toe-in or toe-out. Many times simple contact with another car

Track Tuning & Adjustment

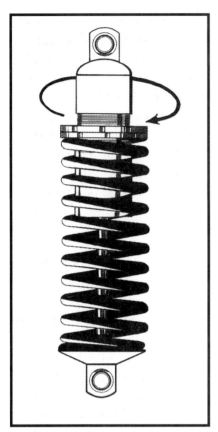

*Putting turns down (clockwise) on the coil-over weight adjusting collar adds weight and corner height to that corner of the car. It **does not** change the spring rate.*

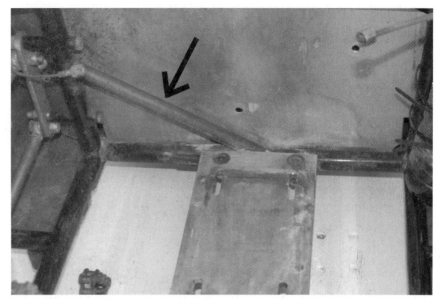

This chassis was a consistent winner for three years, then it stopped being responsive to any chassis adjustments. Upon inspection, several cracks in the chassis were found. The car was returned to the manufacturer for repairs, and two extra braces were added in the center section of the chassis.

will bend something in the steering linkage. If you don't catch this, the car will have excessive toe-in or toe-out, which will cause tire drag on the straightaways and during cornering.

Chassis Tuning With Weight Jacking

Putting turns down (clockwise) on the coil-over weight adjusting collar adds weight and corner height to that corner of the car. It **does not** change the spring rate. Some people think that adding turns down will stiffen the spring, or taking them out will soften the spring. This is not true! Turning the adjusting collar only affects corner weight and height.

Adding turns with the adjusting collar down (clockwise) at the right front will raise that corner of the car and add weight to that corner. This adjustment will make the car tend toward understeer. It will also add more temperature to the right front tire.

This adjustment will also add weight to the left rear corner, and subtract weight from the left front and right rear. As long as the chassis is rigid, the amount of weight that is changed at each corner will be the same. For example, if you turn down on the right front adjuster and add four pounds to that corner, it will subtract four pounds from the left front and right rear, and add four pounds to the left rear.

The Unresponsive Chassis

There may come a time when the race car is not responsive to the normal chassis adjustments that you make, such as corner heights, Panhard bar heights, right rear spacing distance, etc. When this happens, thoroughly inspect your chassis for cracks. Cracks in the chassis create chassis flex. When the chassis flexes, it will not respond to

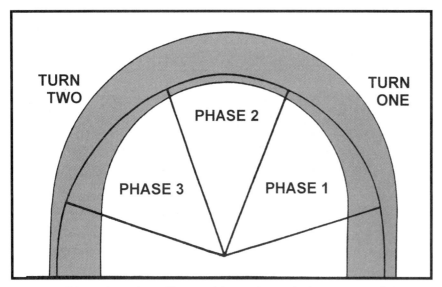

When talking about handling problems through the corners, it is very important to isolate the problem as occurring at one of the three phases of a corner. The three different phases are: 1) corner entry, 2) mid-corner, and 3) corner exit. Each phase of the corner will be affected by what happens at the previous one, so it is important to understand this relationship.

Notice how the right front is pushing. The rear of the car has better grip and is pushing the car forward at corner exit. This can be corrected by increasing stagger or decreasing cross weight.

any adjustment. The chassis has to be torsionally rigid to be able to respond to tuning adjustments.

If you find cracks in the chassis, return it to the manufacturer for repairs.

Handling Problems & Troubleshooting

This section is a guideline to common handling problems most racers encounter, and how to adjust the chassis to handle them.

When there is a problem with the chassis, the racer has to identify where in the turn it is occurring and under what conditions. This identification and isolation can help lead to the correct solution of the problem.

When talking about handling problems through the corners, it is very important to isolate the problem as occurring at one of the three phases of a corner. The three different phases are: 1) corner entry, 2) mid-corner, and 3) corner exit. Each phase of the corner will be affected by what happens at the previous one, so it is important to understand this relationship.

Adjusting For A Push

The type of adjustment made to the car for a pushing condition depends on where the push takes place. For example, if the car is pushing at corner entry, the car is too tight at the right rear. Space the right rear wheel out, or take some right rear weight out of the car. If the car is pushing at corner exit, the car may not have enough stagger. Add more stagger, or lower the front ride height.

Car Is Loose At Corner Exit

This is indicative of having too much stagger in the car.

Decrease stagger. Other things that help here are to raise the front ride height, increase right front and left rear corner weights, move the right rear tire in, or use a softer right rear tire compound.

Bicycling

A car that bicycles (plants the right rear tire and wants to roll over) is generally caused by one of the following problems:

1) The right rear wheel is spaced in too far which causes the right rear to stick too much. Spacing the right rear out widens the rear track and will loosen up the car, and lessens traction on the right rear tire.

2) There is too much weight on the right rear.

3) The rear Panhard bar is set too low.

Tuning For More Grip

When you are battling a slippery race track, there are several adjustments you can make to increase grip:

1) Use a softer compound tire on the right front and right rear (use the same compound at both positions).

2) Take tilt out of the chassis. Less tilt tightens up a chassis.

3) Lower the front and rear Panhard bars an equal amount. This lowers the roll centers and increases lateral grip on the outside tires.

4) Go one step softer on both the right front and right rear springs. Make the spring changes an equal amount so the roll couple distribution does not change from front to rear. Softer springs all more chassis roll which will improve lateral grip.

Tuning For Less Grip

There are situations when you need to reduce lateral grip. One situation is when your car is running in a lower powered class. Too much tire grip (getting the car hooked up too well) will cause these cars to bog down when accelerating off the corners, which slows the car. Another situation that all classes can experience is when there is a lot of tire rubber build-up on the track. This adds significant tire grip. Again, this causes a car to bog down during acceleration off the corners.

When a car is loose at corner exit, it may have too much rear stagger. Other adjustments that may help include increasing cross weight, or moving the right rear tire in.

To reduce grip:

1) Use a harder compound tire on the right front and right rear (use the same compound at both positions).

2) Add tilt to the chassis. More tilt loosens up a chassis.

3) Raise the front and rear Panhard bars an equal amount. This raises the roll centers and reduces lateral grip on the outside tires.

4) Go one step stiffer on both the right front and right rear springs. Make the spring changes an equal amount so the roll couple distribution does not change front to rear. For example, go 10 pounds per inch stiffer on both the right front and right rear springs. Using stiffer outside springs is only effective on smooth tracks. On rougher tracks, softer spring rates are required to ensure tire compli-

Using tire temperatures can give you meaningful feedback about the handling of the race car. For example, temperatures from this car would show that the right front tire has too much negative camber, and the left front doesn't have enough positive camber.

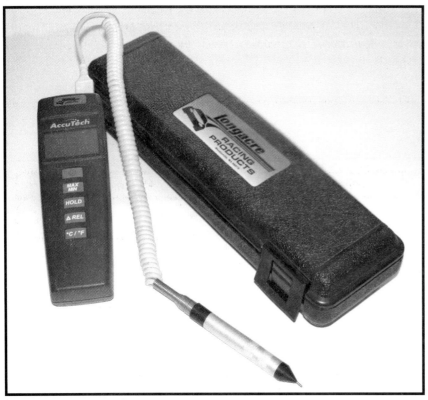

The probe style of tire pyrometer is more accurate than a heat gun for taking tire temperatures.

ance with the track surface.

Chassis Tuning With Tire Temperatures

Because all suspension adjustments and improvements are for the benefit of improving the grip of the tire on the track surface, it stands to reason that tire temperatures are the best indicator of what the tires and suspension system are doing. In fact, the tire temperature method is the only chassis tuning method where it is possible to get away from guessing and work with scientific accuracy.

The instrument used to measure tire temperatures is called a tire pyrometer. There are two types of pyrometers – a probe type, and an infrared type (commonly called a heat gun). The probe type has a needle-type tip which is inserted into the tire rubber through the surface and into the tire cord. The infrared style pyrometer measures the reflective heat coming off of the tire surface.

The probe style pyrometer is much more accurate. The surface temperature of the tire can cool dramatically by the time you get to measure the tire temperatures. On the other hand, the tire heat at the cord remains longer relative to the operation of the tire on the track surface, and presents a much more accurate reading of tire heat. The only drawback to using the probe type of pyrometer is that it takes longer to collect the data than when using the heat gun.

Tire temperatures can change very quickly and cool off fast. Take the temperatures very quickly, starting at the right front and working around the car in a circular motion to the right rear, ending up at the left front. When using the probe type of pyrometer, give the needle time to warm up and stabilize at the first insertion, then don't let it cool off when going from one tire to the next.

Tire temperatures are read at three positions across the face of each tire and are recorded systematically on a sheet of paper in the manner shown in the accompanying example.

By comparing the temperatures across the face of each front tire, it can be determined if each tire has too much or

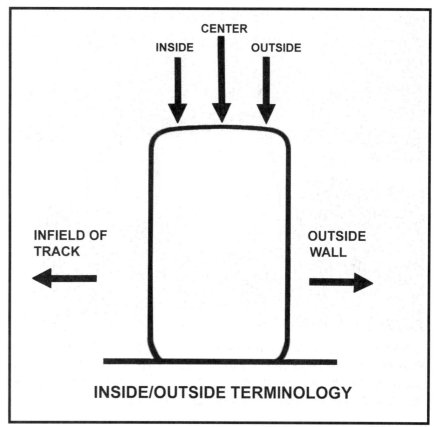

Tire temperatures are read at three positions across the face of the tire, as indicated by the top three arrows. When using a probe style pyrometer, insert the needle at a 45-degree angle to the tire face, not straight in.

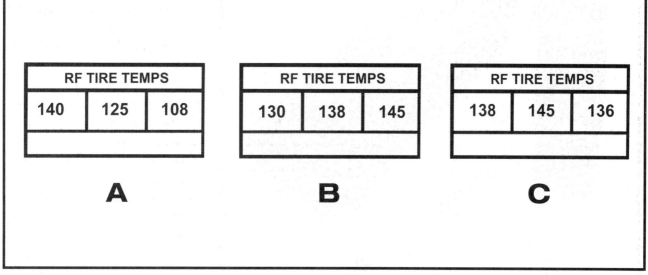

In example A, the inside temperature is much warmer than the other temperatures, which indicates too much negative camber. In example B, the tire has too much positive camber. In example C, the center temperature is the warmest, which indicates too much inflation pressure.

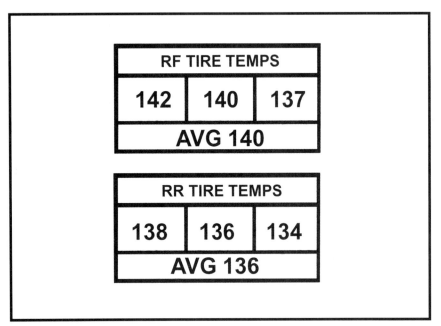

The right side average tire temperatures should be within five degrees of each other. This establishes a front to rear neutral handling balance. These temperatures show that this would be a well balanced car.

not enough camber, or if the tire inflation pressure is correct.

Comparing the average temperature of all three positions of the right front tire to the average of the right rear tire will determine if the chassis is tending toward understeer or oversteer.

The right side average tire temperatures should be within five degrees of each other. This establishes a front to rear neutral handling balance.

Chassis Adjustment Quick Reference

The charts on the following page give a quick reference guide on how to adjust the chassis for oversteer (loose) and understeer (pushing) conditions. They are divided into categories for unlocked rear axle cars and locked rear axle cars. The chassis adjustments are listed in order of the easiest change to the more time-consuming changes.

Adjusting the Chassis For A Dirt Track

There are many chassis tuning elements available on the race car to alter the handling for particular dirt track conditions. Spring rates, shock absorber damping, corner heights, tire pressure, stagger and wheel tracking are some of the elements that can be changed to adapt the chassis to the track and improve handling. The quick reference charts that follow list those adjustments, looking at what can be changed for a sticky track and a dry slick track so you can understand the types of adjustments necessary for these two extremes of track conditions.

Slick Dirt Track

The problem on a slick track is that there is very little traction available from the track. The car is slipping and sliding. The tires offer little grip on the track surface. Chassis adjustments have to be made that tighten up the chassis and help the car drive straight. You need to transfer weight onto the right rear at corner entry to provide side bite and tighten the chassis.

Sticky Dirt Track

On a sticky track, the car has very good traction all around the track, so you have to get the car loosened up to get it to turn and prevent it from pushing. The track has too much traction, so the chassis has to be loosened up. You need to loosen up the car to get it to turn at corner entry. At mid corner and corner exit, the left front corner will rise up under acceleration and load the right rear corner, which will cause the car to push.

Reading A Dirt Track

The most important part of dealing with handling adjustments on a dirt track is knowing how to read a dirt track,

Chassis Adjustment Quick Reference

Car Is Understeering – Unlocked Rear Axle
1) Move right rear tire out
2) Take out weight at right front coil-over
3) Add weight at right rear coil-over
4) Increase front axle lead
5) Raise entire rear of car
6) Use softer tire compound at right front
7) Decrease right front spring rate
8) Increase right rear spring rate
9) Raise rear Panhard bar on frame side

Car Is Oversteering – Unlocked Rear Axle
1) Move right rear tire in
2) Add weight at right front coil-over
3) Take out weight at right rear coil-over
4) Reduce front axle lead
5) Lower entire rear of car
6) Use softer tire compound at right rear
7) Increase right front spring rate*
8) Decrease right rear spring rate
9) Lower rear Panhard bar on frame side

*Helps on corner entry and middle, but not at exit

Car Is Understeering – Locked Rear Axle
1) Move right rear tire out
2) Take out cross weight at left rear coil-over (lower left rear corner)*
3) Increase rear stagger
4) Decrease left rear tire pressure (this decreases cross weight and increases stagger)
5) Increase front axle lead
6) Use softer tire compound at right front
7) Raise entire rear of car
8) Increase right rear spring rate
9) Decrease right front spring rate**
10) Decrease left rear spring rate
11) Raise rear Panhard bar on frame side

*Cross weight is effective at mid-corner and corner exit
**Decreased right front spring rate is effective at corner entry and mid-corner

Car Is Oversteering – Locked Rear Axle
1) Move right rear tire in
2) Add cross weight at left rear coil-over (raise left rear corner height)*
3) Decrease rear tire stagger
4) Increase left rear tire pressure (this increases cross weight and decreases stagger)
5) Decrease front axle lead
6) Use softer tire compound at right rear
7) Lower entire rear of car
8) Increase right front spring rate**
9) Decrease right rear spring rate
10) Increase left rear spring rate
11) Lower rear Panhard bar on frame side

*Increased cross weight helps only at mid-corner and corner exit
**Increased right front spring rate helps only at corner entry and mid-corner

Dry Slick Track Adjustments

1) Raise ride height at all 4 corners
2) Move right rear wheel in
3) Increase cross weight
4) Use a stiffer spring rate at the left front. This stops the car from dropping onto the left front corner at entry, and keeps more weight on the right rear.
5) At mid-corner you want the left front to rise up and transfer weight quickly to the rear. Use a shock with a soft rebound damping. So, a split valve shock (which is stiffer in compression and softer in rebound) is helpful at the left front to keep the corner from dropping down at entry and to help weight transfer quickly to the right rear at exit.
6) Decrease rear stagger.
7) Move the left rear wheel out away from the car. Moving the left rear out decreases the amount of left rear weight and increases right rear weight.
8) Decrease left rear tire pressure to increase the tire patch size.
9) Lower the right rear air pressure to tighten the chassis. Less air pressure increases the tire patch size.

Sticky Track Adjustments

1) Increase the tire air pressure in all four tires. This decreases tire footprint size and tire grip.
2) Use a greater amount of stagger.
3) Move the right rear wheel out away from the car. This takes weight off the right rear corner.
4) Lower the left front corner. This helps the car drop down on the left front at corner entry, which will take weight off the right rear.
5) Use a softer spring rate at the left front. This helps the left front corner drop down easier at corner entry and unloads more weight from the right rear. With less weight on the right rear, the car will turn in easier.
6) Lower the left rear corner. With the corner lowered, the chassis will have less weight on the right rear corner, taking away from side bite and allowing the car to enter the turn easier.
7) Use a tie-down shock on the left rear corner. This shock has softer valving in compression and stiffer valving in rebound. The heavier rebound restricts weight transfer to the right rear at corner entry, which loosens up the chassis.
8) Raise the right rear corner up. This puts more tilt in the chassis, which will loosen the chassis.
9) Use a right rear spring that is one step stiffer. This will loosen the chassis.

and interpret clues about how the track will change.

Outside temperature, humidity and the type of soil all influence how a dirt track changes during a day of racing. How the track operator has prepared and watered the track also has a major influence.

Anticipating the track conditions – and how to dial-in the chassis for them – is the key to keeping the tires hooked up and going fast.

As the racing day progresses, walk the track and observe it. Watch the other cars and how they are hooking up, or are having problems. Look for ruts and bumps and holes that are developing. When you walk the track between races, pick up a piece of loose dirt from the cushion and roll it around in your hand. Feel how well it sticks together. If it just falls apart in your hand, chances are the track is going to get slick. The dirt isn't sticky enough or moist enough to stick together. If it stays stuck together or is gummy, chances are there is enough moisture in it to help the tires get a good bite for awhile.

Kick the dirt with your heel or dig a screwdriver in it to see how hard it is. This gives an indication of whether or not it is going to pack down hard and get slick. If the top surface is wet or has some moisture, dig down deeper to see how deep the moisture goes. If the screwdriver only goes down an inch or so, it is a good indication the track will dry out real quick. If you can push the screwdriver in all the way, it indicates the track will probably hold moisture.

Did the track operator just water the track? You have to watch how much water was put down on the track and determine if it has penetrated down into the dirt. Sometimes the track is dry before the main event and they water it. That might give you a slippery slimy top surface over a hard track, or it might produce a nice tacky top surface. If so, the good tacky bite isn't going too last long. The moist dirt will get scraped off the top layer, and you will have a hard, dry slick surface for the main event. What is important is predicting the condition of the track in the last third of the race, and tuning the chassis to that condition.

Feel the track temperature – is it cold, or warm, or hot? Track temperature will have an influence on the tire compound to run. A track that is cold will allow a softer compound. A track that is hot is going to wear down tires faster, and will require a harder compound. How much harder will depend on how abrasive the track is.

Outside air temperature also has an influence. Real hot ambient air temperature can evaporate moisture out of a track, so even if it starts out real tacky, it can be dry slick for the main event. A cooler or more humid condition will help keep more moisture in the track.

The chassis has to be set up for the end of the race – not the beginning. Almost all tracks are going to get slicker and harder toward the end of the race, so you have to anticipate the conditions and have the chassis set up to work best under those conditions. If you anticipate the track getting harder and slicker, you will have to start the race with the chassis tighter than you would like it to be. But, you will get faster toward the end of the race when the track comes to your setup.

Watching the track and how it progresses during the day will give you an insight as to how hard or how dry the track will be toward the end, and then give you an indication of how much stagger to use, how high to set the front and rear of the car, how to set wheel tracking, and which tire compound to choose.

Because of changing track conditions since the last time your car was on the track, you may have to make several changes to the car at one time. You will have to be able to anticipate how these multiple changes will make the car feel. You want the car to respond in a predictable manner.

Chassis Changes Required For Dirt Tracks

The following items discuss how these chassis adjustment tools work to change the car's handling for particular situations.

Spring Rates

Dirt track spring rate applications are usually 15 to 20 pounds per inch softer than paved track springs for the same track configuration. Spring rates are softer so that increased chassis roll and overturning moment produce more lateral tire grip on the right side tires. The spring rates are also softer so that they are better able to handle track surface bumps, ruts and irregularities.

Spring rates can be adjusted to suit the type of dirt surface being raced on. In general, softer spring rates work best on dry slick tracks, and stiffer spring rates work best on heavier tracks.

Tracking

Moving the right rear wheel out frees up the chassis when the track is rough and sticky. Moving the right rear in when the track gets slick tightens up the car.

A heavy or tacky track will require more stagger. A dry, slick track will require less stagger.

Tilt

Chassis tilt is changed as stagger and right rear air pressure are adjusted. Adding stagger and right rear tire pressure adds tilt, which loosens the chassis. This is desirable for heavy or tacky tracks. Reducing stagger and right rear air pressure reduces tilt, which is desirable for dry slick tracks.

Tire Pressure on Dirt

Tire pressures are varied depending on how much adhesion there is in the track. If a track has a lot of adhesion, such as a tacky track, it requires higher tire pressures. This makes the tire contact patch smaller and lessens the tire grip on the track surface. A higher tire pressure will loosen up the chassis, especially when used at the right rear.

As tire grip on the track is reduced, the tire pressure is reduced to enhance traction. Tire grip is improved because the lower inflation pressure creates a larger footprint. Adjust inflation pressure in one PSI increments.

Chassis Ride Height

Raising the chassis ride height raises the center of gravity of the car. This creates more weight transfer, both front to rear and left to right during cornering. Adding turns in at all four of the coil-over adjusters will raise the chassis height. Doing this will help to tighten the chassis through all three phases of cornering on a slick track.

Chassis Baseline Setup

Left Front

Tire_____
Compound_____
Circumference_____
Air Pressure_____
Spring Rate_____
Shock_____
Ride Ht._____
Camber_____
Caster_____
Corner Weight_____
Panhard Bar Ht._____

Date_____
Car_____
Class_____
Track_____

Right Front

Tire_____
Compound_____
Circumference_____
Air Pressure_____
Spring Rate_____
Shock_____
Ride Ht._____
Camber_____
Caster_____
Corner Weight_____
Panhard Bar Ht._____

FRONT

Front Axle Lead_____

Length_____   Length_____

Rear Axle Lag_____

REAR

Left Rear

Tire_____
Compound_____
Circumference_____
Air Pressure_____
Spring Rate_____
Shock_____
Ride Ht._____
Corner Weight_____
Panhard Bar Ht._____

Locked_____
Unlocked___

Weight Percentages

Left %_____
Rear %_____
Cross Wt.____

Right Rear

Tire_____
Compound_____
Circumference_____
Air Pressure_____
Spring Rate_____
Shock_____
Ride Ht._____
Corner Weight_____
Panhard Bar Ht._____

Gears_____

Suppliers Directory

American Power Sports
12300 Kinsman Rd
Newbury, OH 44065
(440) 564-8100
www.apskarting.com
    Wheels, tires, accessories

A & J Racing Co.
19917 E. Valley Blvd.
Walnut, CA 91789
(909) 598-9902
www.aandjracing.com
    Complete race cars

Boomerang Racing
4750 Mission Blvd., #C & D
Ontario, CA 91762
(909) 627-2219
www.boomerangracing.com
    Complete race cars

Bubba MotorSports, Inc.
2724 Schuylkill Rd.
Parkerford, PA 19457
(610) 495-2220
www.bubbamotorsports.com
    Quarter midget parts and accessories

Bull Rider Race Cars
3160 S. Valley Rd
Crystal Spring, PA 15536
(814) 735-3561
www.bullriderracecars.com
    Complete race cars

Cheetah Chassis
151 S. Main St.
Danielson, CT 06239
(860) 774-5553
www.cheetahchassis.com
    Complete race cars

Circle Racing Wheels
16918 Gridley Place
Cerritos, CA 90703
(866) 865-5278
www.circlekarting.com
    Racing wheels

Comet Kart Sales
2650 W Main St.
Greenfield, IN 46140
(317) 462-3413
www.cometkartsales.com
    Quarter midget parts and accessories

Fast Track Race Cars
16904 85th Ave E.
Puyallup, WA 98375
(253) 864-4445
www.ftracingforkids.com
    Complete race cars

GT American Race Cars
45383 Industrial Place, Unit 5
Fremont, CA 94538
(510) 657-5054
www.GTAmerican.com
    Complete race cars

Harper Racing, Inc.
661 S. New Middletown Rd.
Media, PA 19063
(610) 566-1146
www.harperracing.com
    Quarter midget parts and accessories

Hunter Mfg
P O Box 790
Sumner, WA 98390
(253) 863-7771
www.huntershocks.com
    Shock absorbers

Lightning Chassis
1523 Allgood Rd
Marietta, GA 30062
(770) 527-4427, 770-971-6480
www.tlchassis.com
    Complete race cars

Longacre Racing Products
16892 146th St SE
Monroe, WA 98272
(360) 453-2030
www.longaceracing.com
    Electronic wheel scales, setup equipment

Suppliers Directory

Magnum Chassis
1340 Addison Dr.
Reynoldsburg, OH 43068
(614) 554-2338
www.team-magnum.com
    Complete race cars

NC Chassis Co. (formerly Nervo-Coggin)
409 Munroe Falls Rd
Tallmadge, OH 44278
(330) 798-7744
www.ncchassisco.com
    Complete race cars

Outlaw Chassis
3113 S. Ridgewood Ave.
Edgewater, FL 32141
(386) 427-8522
www.quartermidget.biz
    Complete race cars

Prodorutti Quarter Midget Supply
(215) 362-0113
www.prodorutti-supply.com
    Quarter midget parts and accessories

Rice Cars
20922 Locust St.
Hayward, CA 94541
(510) 582-5897
www.ricecars.com
    Complete race cars

Scott Rider Motorsports
1212 Baseline Rd.
Ohio, IL 61349
(815) 228-9204
www.rider-inc.com
    Ratchet hubs, hubs, rotors, titanium gear system

Robbie Stanley Racing Inc.
7595 Flowes Store Rd.
Concord, NC 28025
(704) 782-6539
www.robbiestanleyracing.com
    Complete race cars

Storm Chassis
P O Box 359
Nazareth, PA 18064
(610) 759-6383
www.stormchassis.homestead.com
    Complete race cars

Tad Fiser Race Cars
1980 S. Navajo St
Denver, CO 80223
(303) 922-9295
www.tfracing.com
    Complete race cars

Tanner Racing Products
A Div. of Mittler Brothers Machine & Tool Inc.
121 E. Mulberry St.
Foristell, MO 63348
(800) 467-2464
www.tannerracing.com
    Coil-over shocks, springs

TS Racing, Inc.
123 West Seminole Ave.
Bushnell, Fl 33513
(352) 793-9600
www.tsracing.com
    Wheels, tires, accessories

Trimble Motorsports
3620 Charter Park Dr
San Jose, CA 95136
(408) 264-6797
www.trimblemotorsports.com
    Shock absorbers, suspension parts

Zero Error Racing Inc.
251 Wheeler St., #111
Sharon, PA 16146
(724) 346-5898
www.zero-error.com
    Suspension & driveline components

Tech Software, Videos, Books

STEVE SMITH AUTOSPORTS

The Complete Karting Guide
For every class from beginner to enduro. Covers: Buying a kart/equipment, Setting up a kart —tires, weight distrib., tire/plug readings, exhaust tuning, gearing, aerodynamics, engine care, etc. #SI40...$21.95

4-Cycle Karting Technology
Covers: Engine bldg. & blueprinting, camshafts, carb jetting & fine tuning, fuel systems, ignition, gear ratios, aerodynamics, handling & adjustment, front end alignment, tire stagger. #S163...$21.95

Kart Chassis Setup Technology
How a kart chassis works & how to improve it. Understanding a kart chassis, chassis & handling dynamics, front end geometry, chassis baseline setup, tires, track tuning & adjustment, dirt track setup. For 2 & 4 cycles, asphalt & dirt. #S287...$24.95

Kart Driving Techniques
By Jim Hall II. A world-class driver and instructor teaches you the fundamentals of going fast in any type of kart. includes: kart cornering dynamics, learning the proper line, learning braking finesse, how to master trail braking, racing & passing strategies, getting out of trouble, & much more. #S274...$21.95

Beginner's Package

1) Building A Street Stock Step-By-Step
A complete guide for entry level stock class, from buying a car and roll cage kit to mounting the cage, stripping the car, prep tips, etc. #S144...$17.95

2) The Stock Car Racing Chassis Performance Handling Basics
Understanding the fundamentals of race car setup & suspension function makes chassis tuning easier. Explains every concept you need to know about what makes a race car handle.. #S30I...$16.95

3) Short Track Driving Techniques
By Butch Miller, ASA champ. Includes: Basic competition driving, Tips for driving traffic, Developing a smooth & consistent style, Defensive driving tactics, and more. For new and experienced drivers alike. #S165...$17.95

Special Offer: Buy all 3 books above together and the special package price is only **$45.95**. Shipping charge is $10.

The NEW Racer's Tax Guide
How to LEGALLY subtract your racing costs from your income tax by running your operation as a business. Step-by-step, how to do everything correctly. An alternate form of funding your racing. #S217...$19.95

Advanced Race Car Suspension
The latest tech info about race car chassis design, setup and development, Weight transfer, suspension and steering geometry, calculating spring/shock rates, vehicle balance, chassis rigidity, banked track corrections, skid pad testing. #SI05...$18.95 WorkBook for above book: #WB5..$12.95

Racer's Guide To Fabricating Shop Eqpment
An engine stand, hydraulic press, engine hoist, sheet metal brake and motorized flame cutter all for under $500. Step-by-step instructions. #S145...$19.95

Sprint Car Chassis Technology
By Jimmy Sills & Steve Smith Detailed info on chassis setup and alignment • Birdcage timing • Front end geometry • Choosing torsion bar rates • Chassis tuning with shocks • Using stagger and tracking • Tire selection, grooving, siping, reading • Blocking, tilt, rake, weight distrib. • Track tuning • Wings. #S282...$29.95 Special Package Price for Sprint Car Chassis Setup Video & book is $62.95

Bldg. The Pro Stock/Late Model Sportsman
Uses 1970-81 Camaro front stub with fabricated perimeter frame. How to design ideal roll centers & camber change, fabricating the car. Flat & banked track setups, rear suspensions for dirt & asphalt, track tuning. #S157...$24.95 Special package price for book & Bldg. The Stock Stub Race Car Video is only **$57.95**

The Racer's Math Handbook
An easily understood guide. Explains: Use of basic formulas, transmission & rear end math, engine, chassis & handling formulas etc. Includes computer program with every formula in the book & more. #S193C...$29.95

How To Run A Successful Race Team
Covers key topics such as team organization and structure, budgetIng & finance, mngmt, team image, goal setting, scheduling & more. #S265...$17.95

Sheet Metal Handbook
How to form aluminum, steel & stainless. Includes: equipment, bldg. interiors, layout design & pattern-making, metal shaping, & more. #SI90...$17.95

Welder's Handbook
Weld like a pro! Covers gas, arc, MIG, TIG, & plasma-arc welding & cutting. Weld everything from mild steel to chrome moly, aluminum, magnesium & titanium. #S179...$17.95

Get Sponsored
How to create a sponsorship program & proposal with a 7-step approach. A guide to finding prospects, sample correspondence, contract agreement. #S260...$59.95

The Great Money Hunt
This book is ONLY for those who are really SERIOUS about obtaining sponsorship. Provides strategy with proposals, justifications & sample correspondence. How to find prospects, & sell and fulfill sponsor's marketing needs. #S200...$69.95

Selection & Application of Late Model Dirt Racing Tires
If you are serious about maximizing tire traction & properly managing tires, you need this book. A complete analysis of Hoosier's tread compounds & their use, analyzing track soil & conditions, grooving & siping, adjusting air pressures, & much more. #S288 - $29.95

Computer Programs

Computerized Chassis Set-Up
This WILL set-up your race car! Selecting optimum spring rates, cross weight and weight dist., computing CGI'I, front roll centers, stagger, ideal ballast location, final gearing & more. Windows format, CD-ROM. #C155... $115.95

Race Car Simulator-2
Shows effects of chassis changes. Specify chassis & track specs, then the computer shows how the car will handle. Change springs or weight and the computer tells the new handling. Windows format, CD-ROM #C247...$115.95

Race Driving Simulator
This program teaches competitors how to choose a proper racing line based on the traction cirle. Learn: when to apex early or late, when to sacrifice speed to get to the throttle sooner, how to alter your line to pass an equal car, and much, much more. #C293...$129.95

Front Suspension Geometry PRO
Does 3D analysis & graphs all suspension movements Including camber curve, caster curve and bump steer. Makes suggestions on what parts to move & where to improve suspension geometry. For all cars. Windows format, on CD-ROM. #C249...$119.95

Rear Suspension Geometry PRO
Analyzes rear axle bump steer and roll steer of 3-link and 4-bar (aftermarket type) rear suspensions. Shows the amount of rear axle lateral movement, pinion angle change, axle steer and rear tire fore/aft movement. Windows format, on CD-ROM. #C264...$115.95

Stock Car Dirt Track Technology
By Steve Smith Includes: Dirt track chassis set-up & adjustment, rear suspension systems (fifth-coil/torque arm, 4-link, leaf springs), front suspension, shocks, tires, braking, scaling, track tuning, prep tips, & much more. #S196...$29.95

Street Stock Chassis Technology
Chassis & roll cage fabrication, front suspension alignment, changing the roll center & camber curve, rear suspensions, springs & shocks, stagger, gearing, set-ups for asphalt & dirt, wt adjustments, track tuning, & more! #SI92...$29.95 Special Package Price for this book & Street Stock Chassis Setup Video is **$62.95**

Trackside Chassis Tuning Guide
A handy, concise, indexed guide to analyzing and correcting race car handling problems. Organized by problem (such as corner entry understeer, etc.) with proven solutions. Almost every solution to every chassis problem is contained. 3.5"x7.5 #S283...$16.95

I.M.C.A. Modified Racing Technology
Covers: chassis structure & design, step-by-step chassis setup & alignment, chassis tuning with shocks, track tuning, front suspension, all rear suspensions - 3-link, swing arm, 4-link, coil/monoleaf, & more. Everything to get you competitive. #S280...$29.95 Special Package Price for book & Building The IMCA Video is only **$62.95**

Building the Mustang Mini Stock
Stripping the car, installing a roll cage, working with a unibody chassis, McPherson strut front suspension, 4-point rear suspension and coil springs, complete car fabrication, high performance engine buildup, chassis setup. For dirt & paved tracks. #S289 - $29.95

Paved Track Stock Car Technology
Step-by-step chassis setup & alignment • Front suspension & steering • Rear susp. systems • Selecting springs & shocks • Chassis tuning with shocks. Scaling the car • Track chassis tuning . #S239...$29.95

Pony Stock/Mini Stock Racing Technology
Includes: Complete performance build-up of the Ford 2300 cc engine • Fabricating & prepping a Pinto chassis • Scaling & adjusting the wt. • Detailed chassis setup specs & procedures. Front & rear suspensions • Low buck part sources • Track tuning. For dirt & paved tracks. Covers Pro-4's too. #S258...$29.95

Dwarf Car Technology
Chapters include: Performance handling basics, front suspension and steering, rear suspension and driveline, shock absorbers, tires and wheels, the braking system, chassis setup in the shop, adjusting handling to track conditions, and safety systems. #S225 - $24.95

Minisprint/Micro Midget Chassis Technology
Complete chassis setup, alignment and blocking, chain drive systems, gearing, front and rear suspension systems, springs and torsion bars, tires and wheels, using shocks to fine-tune the chassis, adjusting the chassis to changing track conditions, identifying and solving handling problems, wings and aerodynamics. For all 125, 250, and 600 micros, and 1000 to 1200 minisprints. #S286 - $29.95

Expert Chassis Set Up Videos

Chassis Setup with Tire Temps	#V232...$29.95
Building the I.M.C.A. Modified	#V181...$39.95
Building The Mustang Ministock	#V290...$39.95
Paved Track L.M. Chassis Setup	#V279...$39.95
Dirt Late Model Chassis Setup	#V261...$39.95
Dwarf Car Chassis Setup	#V281...$39.95
Midget Chassis Setup	#V194...$39.95
Sprint Car Chassis Setup	#V182...$39.95

Order Hotline 714-639-7681
Fax 714-639-9741

Steve Smith Autosports
P O Box 11631-W
Santa Ana, CA 92711

www.SteveSmithAutosports.com

S&H: Add $7 for first item, $8 for 2 items, $10 for 3 or more